THE ANTHROPOLOGY OF PRECIOUS MINERALS

The Anthropology
of Precious Minerals

EDITED BY ELIZABETH FERRY, ANNABEL
VALLARD, AND ANDREW WALSH

UNIVERSITY OF TORONTO PRESS
Toronto Buffalo London

ISBN 978-1-4875-0317-8

Library and Archives Canada Cataloguing in Publication

Title: The anthropology of precious minerals / edited by Elizabeth Ferry,
    Annabel Vallard, and Andrew Walsh.
Names: Ferry, Elizabeth Emma, editor. | Vallard, Annabel, editor. | Walsh,
    Andrew, 1969–, editor.
Description: Includes bibliographical references and index.
Identifiers: Canadiana 20190181516 | ISBN 9781487503178 (hardcover)
Subjects: LCSH: Precious metals – Social aspects. | LCSH: Mineral
    industries – Social aspects. | LCSH: Mines and mineral resources – Social
    aspects. | LCSH: Metal-work – Social aspects. | LCSH: Anthropology.
Classification: LCC GN436 .A58 2019 | DDC 306.4/6 – dc23

University of Toronto Press acknowledges the financial assistance of the
Wenner-Gren Foundation for Anthropological Research in the publication of
this book.

University of Toronto Press acknowledges the financial assistance to its
publishing program of the Canada Council for the Arts and the Ontario Arts
Council, an agency of the Government of Ontario.

 Canada Council for the Arts Conseil des Arts du Canada

 ONTARIO ARTS COUNCIL
CONSEIL DES ARTS DE L'ONTARIO
an Ontario government agency
un organisme du gouvernement de l'Ontario

Funded by the Government of Canada Financé par le gouvernement du Canada

Contents

Part Two: Mineral Connections

Acknowledgments

This book is based on a workshop sponsored by the Wenner-Gren Foundation that was held at the Royal Ontario Museum in Toronto, Canada. We would like to thank the Wenner-Gren Foundation for Anthropological Research for supporting this project from beginning to end and, especially, Laurie Obbink for her patience and encouragement. We would also like to thank colleagues at the Royal Ontario Museum for making our time there so productive; we are especially grateful to Sarah Fee for her considerable help in planning and putting on the workshop and to Katherine Dunnell for help in planning the hands-on components of the workshop, to other ROM staff for introducing us to so many of the museum's precious specimens, and to Sylvia Forni and Chen Shen for additional support. We would also like to thank the University of Western Ontario's Faculty of Social Science for additional financial and administrative support, Marie-Pier Cantin for painstaking editorial work, and Jeannie Taylor for great help in administering the finances of the project. Thanks also to Richard Hughes for recommending sources and to Lindsay Bell and Brian Brazeal, whose contributions to these discussions and ongoing work have helped to shape this volume. Finally, thanks to colleagues at the University of Toronto Press (Doug Hildebrand, Jodi Lewchuck, and Janice Evans), to copy editor Carolyn Zapf, to indexer Stephen Ullstrom, and to the reviewers of earlier drafts of this volume.

THE ANTHROPOLOGY OF PRECIOUS MINERALS

Introduction: Engaging Precious Minerals

ANDREW WALSH, ELIZABETH FERRY,
AND ANNABEL VALLARD

30 April 2015. Midday. Following a morning of presentations, a dozen visitors tour the Royal Ontario Museum's newly renovated Teck Suite of Mineral Galleries. Like students on a class trip, they follow their guide, ROM mineralogist Katherine Dunnell, as she talks them through the spectacular diversity of the collection. An hour later, they gather in a closed meeting room to engage more intimately with the stuff that's brought them here: an array of specimens taken out of storage for them to hold, to eye up close, to pass around. Everyone knows how to use the magnifying loupes on hand, and everyone has a story to tell. A stranger entering the room might confuse this social science workshop for a meeting of eager gem lovers, rock hounds, or collectors whose passion for minerals couldn't be contained.

This volume is the product of a workshop in which the moments we just described took place. Sponsored by the Wenner-Gren Foundation for Anthropological Research and held over two days in the spring of 2015 at the Royal Ontario Museum (ROM) in Toronto, this workshop enabled participants to address the complexity of human-mineral engagements through ethnographic case studies and anthropological reflections on research with diverse human actors (cellphone scrappers, mountaineering crystal hunters, artisanal miners, gem traders, bead collectors, lapidaries, and auctioneers, among others), as well as with their various mineral partners (gold, crystal specimens, sapphires, emeralds, diamonds, and so forth). It also gave us the chance to share in the complementary interests of mineral specialists, ROM curators and mineralogists especially, and to engage, unguarded, with minerals themselves.

In past discussions with one another, we, the authors of this introduction, have often remarked on the contagious passions of the mineralogists and other mineral specialists with whom we have come into contact through our anthropological research, and, by extension, on how difficult it can be to put a mineral specimen down once you've got one in your hand (especially if you have a jeweller's loupe in the other) – every turn of the wrist reveals something new. Although researchers who study humans and those who study minerals approach human-mineral engagements from different directions, we are similarly prone, it seems, to taking multiple, careful perspectives on what others take for granted, seeing fundamental processes, wondrous complexity, and, often, beauty manifested in what others might dismiss as mundane. In an early planning meeting with ROM mineralogist Katherine Dunnell, Andrew Walsh was particularly struck by her use of a classic strategy employed in first-year anthropology courses, that is, trying to enrich the understanding of minerals by making the strange familiar and the familiar strange, first pointing out how some of the spectacular mineral specimens on display at the ROM are of the same kinds found in laptops, cellphones, toothpaste, and house paint; and then revealing a secret world of minerals hiding in plain sight, for example, coating the pages of glossy magazines and high-end gem and mineral catalogues.

Our tour of the ROM's mineral gallery revealed workshop participants' personal and shared mineral passions. Each of us was attracted, for reasons of our own, to different collections, some lingering over the showcased mounted specimens that fill the bulk of the gallery, others drawn to a vault-like room at the back in which cut and polished gemstones are displayed behind thicker glass. As anthropologists, however, we were also aware of how the attractive, inspirational, or otherwise affective qualities of these minerals are relative and by no means universally apparent. One handler's sapphire fetish is another's commodity; one's showcased specimen is aggregate for another's concrete. And so, as much as we were drawn in by the spectacular mineral diversity on offer at the ROM, we were also intrigued by the spectacle of the mineral gallery itself and the vision of the world it contained: its representation of the history of mining (including displays devoted to the Canadian Mining Hall of Fame); the role that public and private donors (including a mining company) have played in making this history and its spoils available to museum-goers; and, especially, its efforts at educating visitors on what mineral extraction entails. One memorable part of our tour involved an interactive exhibit located just within the gallery's entrance. Gathered around a large touchscreen table, we were invited to take on the roles of different players in a large-scale mining project – workers, mine managers, and investors, among others – so as to better reckon with the many and complex effects of strikes, economic downturns, and other scenarios common to the mining industry. As we played the game, we wondered: How would other museum-goers play it and to what ends? What would they make of the information relayed through this innovative exhibit alongside the complex messages communicated by the dozens of beautifully lit showcases around it? Would they view the specimens and gems on display in the gallery around them as works of natural art or as condensations of key social material relations, some of which were being played out in the game?

Our time together at the ROM enabled a heightened sense of how human-mineral relations might be approached and understood as both deeply personal and institutionalized, shaped simultaneously by affective, sociotechnical, and political-economic processes. It also bolstered our conviction that human-mineral engagements warrant the attention we were paying them. As the ROM's mineralogists and exhibits reminded us, minerals are quite simply fundamental features of human lives, not to be overlooked, ignored, or imagined solely as museum pieces. The volume that follows builds on this conviction, offering reworked versions of presentations and commentaries delivered at the ROM workshop. To these essays we have added this introduction and an afterword, both informed by our ongoing discussions and readings of each other's work. In sum, we offer cases and reflections that say something about human-mineral engagements in general and, more specifically, about the quality of "preciousness" that commonly emerges from at least some of these engagements.

Relationships among minerals, materiality, and meaning are addressed throughout the chapters that follow, corroborating Ferry's point that, as "the material substrate of materiality itself," minerals are especially "good to think about how value becomes solidified in the material world" (2013: 10). Each in their own way, contributors attend to the specificity of particular minerals – copper, aluminium, iron, and other metals (Joshua A. Bell); sapphires (Andrew Walsh); crystal minerals (Gilles Raveneau); jade, corundum, carnelian, and other hard stones (Annabel Vallard); diamonds (Filipe Calvão); and gold (Les W. Field) – and of what they do to, with, and for humans at particular moments, but also to the varied representations and processes of valuation in which such minerals are caught up on their way to becoming "precious" (to some at least), moving along their singular "biographies" (Kopytoff, 1986) through mining communities, mineral shows, museums, and networks of skilled specialists, traders, and collectors. We would like not only to highlight what is distinctive about the people and communities involved in producing or fulfilling demand for a defined category of minerals but also to consider how a perceived quality like preciousness is itself multifaceted, emerging from different sorts of engagement through which humans and minerals interrelate.

The building that hosted the workshop and the mineral gallery we toured underline the intricate presence of minerals in human lives. A recent addition to the ROM, the Michael Lee-Chin Crystal has been described as "one of the most challenging construction projects in North America for its engineering complexity and innovative methods ... [It is] composed of five interlocking, self-supporting prismatic structures that co-exist but are not attached to the original ROM building, except for the bridges that link them. The crystal is intended to evoke the museum's mission, 'to build bridges of understanding and appreciation for the world's diverse cultures and precious natural environments'" (ROM, 2018). What appealed to some of us was not only the ROM's crystal outgrowth, but also the location of the museum's mineral gallery alongside a biodiversity gallery focusing on "life in crisis," which featured endangered and extinct species that highlight the uncertainty of life on Earth in the Anthropocene and the pressure humans exert on "natural" resources, minerals included. Indeed, the Anthropocene, the geological age defined in terms of human effects on the earth in all its dimensions (animal, vegetable, mineral, meteorological, atmospheric, and so on), can be seen as the ultimate proposition of human-mineral engagements.

In stressing the mutuality of human-mineral relations, we join a growing number of others in advocating the need to attend carefully to the complexity of human engagements with the non-human, whether dogs (Haraway, 2003), forests (Kohn, 2013), microbiomes (Helmreich, 2009), mushrooms (Tsing, 2015), and other organic beings; or stones (Ingold, 2007), volcanoes (Palsson & Swanson, 2016), and other "existents" "beyond the carbon imaginary" (Povinelli, 2016). As such, our work has also been influenced by trends in thinking associated with what some have termed "the ontological turn" in science and technology studies (STS), anthropology, and philosophy (Jensen et al., 2017).[1] We should specify, however, that, while several of the chapters ahead bear this influence in productive ways – for example, highlighting the reciprocal nature of various extractive interventions (Hacking, 2002) and reflecting on the agency of minerals – we do not intend the volume as a whole to be read as an argument for one or another approach in this rapidly developing area of scholarship.

Mineral Affordances and Actions

What can a stone do? A stone can endure, it can change, it can harm, it can heal. It can make you rich, it can make you poor, it can become an enemy, a friend, and a teacher. It can carry your memories and your dreams. It can build empires and bury cities. It can reveal the history of the universe. It can open and close the gates of philosophy. It can change the course of nature. It can change its own nature. It can empty the world of time. (Raffles, 2012: 527)

As Jeffrey Jerome Cohen notes in *Stone: An Ecology of the Inhuman*, there is a "long tradition of mining the philosophical from the lithic" in research and reflection (2015: 4). While such metaphorical mining offers different resources to distinctive pursuits, what motivates Cohen, the contributors to this volume, and so many other prospectors in this vein is an appreciation of the vitality of a mineral world that is too commonly imagined as the passive foundation on which human life plays out. As Cohen reminds us, stone is "[n]either dead matter nor pliant utensil" but a "bluntly impedimental as well as collaborative force" that "brings story into being" (7). Our own realizations and understandings of this force have been inspired by an eclectic array of sources that, as we present them here, suggest the breadth of the long-standing tradition of thinking to which we intend this volume to contribute. Among these sources, works of poetry offer an immediate and inspiring sensory immersion into human-mineral worlds and are thus a great place to begin.

The opening lines of American poet Lorine Niedecker's poem "Lake Superior"[2] underline the intersecting pathways travelled by people and stones:

In every part of every living thing
Is stuff that once was rock

In blood the minerals
of the rock

Iron the common element of earth
in rocks and freighters (Niedecker, 2013: 1)

Based on a trip around the lake that Niedecker took in 1966, the poem opens with a startlingly direct image of human-mineral entanglements. The refusal to respect the boundaries between carbon-based and non–carbon-based life (Povinelli, 2016) and between nature and society brings rocks and minerals to the service of a particular view of the formation of the world, society, "America" as founded through missionizing and resource development, and the land with its rocks and minerals later visited by tourists on their summer vacations. The poem and its accompanying texts throughout the book, also entitled *Lake Superior* (Niedecker, 2013), show that the lake region has been created by its minerals and rocks, as well as by the Native Americans who lived and still live there, and the French and English explorers who wrote about it. The book intersperses excerpts from the writings of the explorers Pierre Esprit Radisson and Henry Rowe Schoolcraft (1661 and 1832, respectively), with descriptions of the agate, hornblende, carnelian, and iron of the region and small asides that anchor us in human time, such as the time of vacation ("Why should we hurry home?"). These texts together produce a picture of Lake Superior country (in which there is little about the lake itself, in fact) that inhabits several times at once – geological, historical, and personal. This conceptual move, bringing multiple temporalities into one frame, is characteristic of contemplations of the mineral world. Minerals make us think about the vastly different time scales within which they, and we, move (Raffles, 2012; Ferry, 2015).

While complex and layered mineral landscapes inspire some poets, particular mineral formations fascinate others, including Roger Caillois,[3] due to their "singularity." Curious, unusual, and amazing, they "catch attention through some anomaly of form, some suggestive oddity of color or pattern" (Caillois, 1985: 1). At a reduced human scale, these spectacular specimens, often valued for their surfaces, lead their amateurs to the depths of the mineral world. Caillois's fascination with the image in the stone, for example, gives him the impression that "nature" is representing something that should be deciphered through imagination and analogical relations to the milieu. In a particular stone, he sees miniaturized ruins, *septaria*, or landscapes that may be held in the palm of his hand for his benefit alone, revealing both the human need to find similarities in things and the intimate familiarity of his own relations with his collection. Throughout the years, Caillois often looked at, handled, and caressed each of these pieces of "natural art" (104), taking inspiration from his sensory/sensuous engagements with mineral specimens for writing projects intended to give his readers an idea of the proportions and laws of the general "beauty" or "aesthetic of the universe," fundamentally external to any human interventions (2–3). Having known some of his specimens well, he gradually gave up regarding humans as external to nature, learning instead to regard humanity as part of a "scale ranging from molecules to the stars" (Yourcenar, 1985: xi), "the tissue of the universe (being) continuous" (Caillois, 1985: 103).

Like Niedecker and Caillois, Francis Ponge (1971) also questioned the multiple ways in which humans dwell and become embedded in a world from which they cannot be separated. The interest of this "poet of things" in minerals stems from his meticulous attempts to represent the experience of everyday objects and to give voice to "the mute world" of things, from objects to matter (276). Unlike Caillois, however, Ponge describes the fundamental processes of this world not with reference to spectacular specimens but through mundane, simple, and silenced things: gravel as well as gemstones. Pebbles, in particular, drew his attention due to the perfection of their forms, so conducive to handling and playing in the poet's hands. Larger than gravel, smaller than rocks, pebbles come, says Ponge, with their own history of taking successive forms from the "grey chaos" (Tcholakian, 1989: 84). While Caillois emphasizes the immutable quality of individual notable specimens, Ponge underlines the lively finitude of minerals. "Contrary to popular opinion, which makes stone in man's eyes a symbol of durability and impassiveness," he wrote in 1942, a pebble that has been fragmented and polished through contact with elements, sand, water, wind, plants, and animals "does not regenerate, is in fact the only thing in nature that constantly dies" (Ponge, 1974: 73). Of no use to humans, the pebble is "still wild, or at least not domesticated," a stone before its significations that humans may experience only momentarily through their perceptual abilities, with their hands, their eyes, their whole body (74).[4] For Ponge, to deal with this pebble and, more generally, with the slightest, most mundane object is an opening "to the complexity of the universe," echoing Jan Zalasiewicz's more recent *The Planet in a Pebble* (2012), a book that tells the story of Earth's deep history with reference to a single pebble picked up on a beach in Wales. In encouraging the contemplation of mundane things – as did Tim Ingold (2007) in inviting readers to pick up and immerse a "largish stone" in water before reading his paper "Materials Against Materiality" – Ponge decentres the anthropocentric perspective, putting "the human in its place in nature" (Ponge, 1948: 225) by way of emphasizing his own sensuous entanglement in a world that is always to *occur* (Ingold, 2007: 14) through relations with the endlessly variable properties of materials, not least the "stoniness" of stones (Tilley, 2004: 220).

Moments of transformation between "organic" and "inorganic" worlds provoke consideration of the place and displacement not only of humans but of all carbon-based forms. Katie Peterson's (2013) "Fossil Necklace" offers a profound reflection on the evanescence of these forms as well as on their inner, fragile continuity. She holds on a thin thread an assemblage of 170 fossils collected and carved into spherical forms, each referring to a founding event related to life on Earth. From single-celled organisms to the first flowering of plants, from the first flight of birds' ancestors to the extinction of American megafauna, these mineralizations of life echo the vitality of the minerals we address in this book.[5]

From an anthropological perspective, much can be made of the fact that minerals afford poets, artists, and others such distinctive opportunities for deep reflection. As Nicole Boivin notes in her introduction to *Soils, Stones and Symbols: Cultural Perceptions of the Mineral World*, "minerals are frequently symbolically meaningful, ritually powerful, and deeply interwoven into not just economic and material, but also social, cosmological, mythical, spiritual and philosophical aspects of life" (Boivin & Owoc, 2004: 2). But minerals have also proven useful to humans and those who study them in more practical ways, not least as the sources of tools with which trajectories of biological evolution and cultural change are often associated (1). Various sorts of stone, clay, copper, iron, and other minerals, as well as the techniques and technologies that have enabled their centrality in human lives, have long figured in making the revolutions and maintaining the relative equilibria by which we have come to understand our species' distinctive history and place in the world. By drawing the material, mineral, bases of stone tools, rock art, monumental architecture, and the like into clearer definition, Boivin and other archaeologists (including Susan D. Gillespie, who introduces this volume's first set of chapters) are envisioning human engagements with the mineral world in new ways, revealing features of the archaeological record that are often "overlooked by [existing] interpretive frameworks" (Boivin & Owoc, 2004: 3). What is more, due to their physical stability and their distinctive chemical and physical signatures (Roberts & Thornton, 2014), flint, clay, obsidian, bronze, tin, gold, jade, lapis lazuli, and other minerals have also served archaeologists as important indicators of early global connectivity, indicating the paths taken by trade routes connecting Asia and Europe, for example, and bearing witness to the various physical and chemical transformations entailed in past *"schèmes opératoires"* (Simondon, 1958) or *"chaînes opératoires"* (Leroi-Gourhan, 1965; Lemonnier, 1976; Balfet, 1991) of extraction, processing, and use (Casanova, 2013; Cauvin et al., 1998).

Like Boivin and others, the contributors to this volume are concerned with how minerals are valued in contextually specific ways that are hard to discern through superficial study – ways influenced by specific processes of extraction, exchange, and valuation, as well as by the distinctive world views and social lives of people who handle and pass them along as commodities, gifts, heirlooms, artefacts, collectibles, samples, specimens, museum pieces, and so on. Like other "things," then, the minerals discussed in the chapters that follow can be understood as having "social lives" (Appadurai, 1986) of their own, prone to being evaluated and re-evaluated as they pass from one "regime of value" to another. Through it all, however, they remain more than what people make of them, retaining what Appadurai elsewhere refers to as the "stubbornness" of "the thing itself," a "chaotic materiality ... that resists the global tendency to make all things instruments of representation, and thus of abstraction and commodification" (2006: 21). The analytical possibilities afforded by this intriguing combination of qualities is beautifully presented in Brian Brazeal's *Illusions in Stone* (2016a), a multi-sited ethnographic film that follows the global pathways of the emerald trade through Brazil, India, Zambia, Tel Aviv, and New York City while remaining firmly focused on what sets emeralds apart from other global commodities.

Our choice to specify the minerals referenced in the coming chapters as *"precious* minerals" is obviously partly informed by conventions of mineralogy, gemology, the jewellery industry, and our native languages (English and French) that would have us classify most of the minerals discussed in this volume as just that.[6] We also see great analytic potential in the notion of "preciousness" itself, however, as a means for capturing important aspects of the human-mineral engagements on which we focus. As Anne-Sophie Trébuchet-Breitwiller (2011) notes in a study of "the work of preciousness" in France's perfume and wine industries, the fact that "preciousness" has long "stuck" to certain kinds of things is not simply a product of repeated applications of a qualifier. Indeed, in the cases she describes, as in the cases collected in this volume, when the term "precious" is invoked with reference to matter at hand (if, in fact, it ever is), it is just one of a list of adjectives that might apply (320). For Trébuchet-Breitwiller, and for us, it is only in attending carefully to discourse around, and observing experiences with, such things that we are pointed to "preciousness" as a keyword that captures something distinctive about people's engagements with them.

Shine, Iridescence, Brilliance

That minerals contribute something important to reckonings of their own preciousness is especially clear when we consider the entanglement of mineralogical, social, and semiotic phenomena and processes underlying the simple fact that "precious" minerals commonly manifest a range of perceptible qualities – shine, sparkle, brilliance, iridescence, and so on – enabled by the interaction of light and specific mineralogical structures.

Sandra Revolon describes how the Owa people of Aoigi, in the Eastern Solomon Islands, distinguish between different forms and degrees of luminosity. Particularly significant are those substances that reflect light rather than produce it and, among them, those that create the optical phenomenon of "iridescence," in which the colour of an object or substance changes with the angle of sight and of the light source. Nautilus, the scales of bonito fish, certain clouds, and the appearance of the ocean at sunrise, sunset, or moonrise can produce this phenomenon; for this reason, Revolon suggests, they have particular connections to the ancestral spirits that watch over living Owa people. Ritual work with these substances at initiation and other times helps to maintain the proper and crucial relation between the living and dead (2012: 252).

Bissera Pentcheva places the brilliance of gilded surfaces in ninth and tenth century Byzantine "relief icons" in the context of other synesthetic aspects:

> In its original setting, the icon performed through its materiality. The radiance of light reflected from the gilded surfaces, the flicker of candles and oil lamps placed before the image, the sweetly fragrant incense, the sounds of prayer and music – these inundated all senses. In saturating the material and sensorial to excess, the experience of the icon led to a transcendence of this very materiality and gave access to the intangible, invisible, and noetic. (2006: 631)

Like Revolon, Pentcheva notes how shininess in its different forms can afford a semiotic vocabulary of power and transcendence, rendering certain kinds of objects "precious." Both authors also exemplify how qualitative affordances for preciousness have been mobilized in different times and places, though not always in the same way or with the same consequences.

In describing "biographies of brilliance" in pre-Columbian American worlds, Nicholas Saunders notes:

> There is a wealth of evidence to suggest that indigenous Amerindians perceived their world as infused with a spiritual brilliance which manifested itself in natural phenomena – sun, moon, water, ice and rainbows; natural materials – minerals, feathers, pearls and shells, and artefacts made from such matter, as well as ceramics, textiles and metals. All in their own way, and according to different cultural conventions, partook of an inner sacredness displayed as surface gleaming. (1999: 245)

For their part, Europeans approached and took the brilliance they encountered in the Americas differently, recontextualizing these broad and varied categories according to their own cultural indices, especially purity and convertibility, and elevating certain substances above others as distinctly valuable. The difference between European and Native American valuations of brilliance highlights the two primary aspects of preciousness generally found in dictionary definitions of the term. While Native Americans valued many brilliant substances in ways that conform to a sense of the precious as something "held in high esteem," Europeans distinguished certain substances such as gold, silver, pearls, and gems as precious both in this sense and also in the sense of being "of great monetary value," a combination apparent in early seventeenth century England, for example, in gold's associations with both royal charisma (through its sun-like brilliance) and monetary value (Wortham, 1996). Indeed, the co-presence of royal power as brilliance and intrinsic worth as money together form a redoubled kind of preciousness linking the Jacobean monarchy and the monetary system, expressed through the stamping of the sovereign's likeness on the golden sovereign coin (351).

These examples highlight how the "preciousness" of certain materials (and what is made from them) can be, like that of perfume or wine (Trébuchet-Breitwiller, 2011), "appraised" and "prized" (Dewey, 1939), not only through alienating processes of abstraction and commodification but also through people's momentary experiences *with* them. Accordingly, we do not understand "preciousness" as an inherent quality of certain minerals (as some classificatory schemes might demand), nor do we think "preciousness" best understood as merely a social construction of those who reckon it in certain minerals. Rather, as noted later and discussed extensively in the afterword, we consider preciousness to be a multifaceted quality that emerges through varied moments and processes of human-mineral engagement.

Extracting Precious Minerals

Anthropological research on the work and business of mining has long emphasized how mineral extraction involves working, skilful, situated people's entanglement with particular minerals, technologies, and sources (landscapes or otherwise). Alex Golub's recent description of Papua New Guinea's Porgera mine as "a complex sociotechnical system made up of a variety of human and nonhuman actors" (2014: 7) can be applied to other contexts of mineral extraction, suggesting rich possibilities for reflecting on how the situated people involved in or excluded by this process make sense of and negotiate contexts in which they never act alone. On the global scale, the cast includes global consumers (Smith, 2011), corporations (Rajak, 2011), and the forces of capitalism itself (Kirsch, 2014). At mineral sources, meanwhile, we encounter another set of participating and potentially responsible agents, sometimes even sites of ongoing or potential extraction themselves (Salas Carreño, 2017; Jorgensen, 1997). Indeed, the closer one gets to sites of mineral extraction, it seems, the more obvious it becomes that this process is always more than an exercise of human ingenuity and labour. Ethnographic accounts of mining work commonly reveal that the greatest advocates of the notion that mineral extraction involves the interplay of human and non-human actors are those closest to mineral sources who have most benefited and/or suffered from this fact: Bolivian tin miners who attribute drops in productivity to "the poor performance of the machine, the low air pressure, the hardness of the rock matrix, or the failure of the engineer to assess practical problems" (Nash, 1993: 178); Wyoming coal miners who share responsibility for accidents with the "capricious" (Rolston, 2013: 591) highwalls they work; Indonesian gold miners who experience the agency of the gold as "luck" (Peluso, 2018: 411); or Mongolian gold miners who are both attracted to and afflicted by "the power of gold" (High, 2017: 53) to give just a few examples.

At the source, minerals commonly matter most today in how they provide incomes and livelihoods for people and, by extension, how they support communities, cooperatives, and other social collectives. As Ferry notes in an account of a silver mining cooperative in Guanajuato, Mexico, "the place where the drill bites into the rock" is the point from which "the life of the cooperative and the livelihood of its members emanate" (2005: 423). Like other sites of extraction considered in this book, Guanajuato's silver mines are sources of both marketable minerals and complex socio-material relations, the mutuality of which are undeniable. What's more, in this context, as elsewhere, it is not only differently positioned people who figure as key players in an unfurling history of extraction. Guanajuato's silver miners have increasingly reckoned as valuable not just silver but calcite, pyrite, acanthite, quartz, and other mineral specimens, enabling these mineral actors' entry into distinctive social lives and relationships, locally and internationally (Ferry, 2005).

Precious minerals are commonly reckoned precious in the markets through which they circulate due to their imagined scarcity. Indeed, it is often vague perceptions of the scarcity of particular minerals that lends urgency to the search for them and allure to the investment opportunities and financial instruments with which they are associated (Tsing, 2000). At their sources, the scarcity of such minerals matters as well, but not only in how they are imagined. Among artisanal miners of gold, diamonds, gemstones, and rare earths, the challenge posed by scarcity is very real. That artisanal mining rushes centred on newly discovered sources inevitably produce far more fruitless workdays than bonanzas says more about the nature of any sort of precious mineral extraction than it does about the inadequacies of prospectors. As Filipe Calvão and Walsh note in their contributions to this volume, contrary to their reputations as anachronistic outcasts of the global economy, artisanal miners are, in fact, often skilled surveyors and workers for whom, as in any industrial operation, getting at precious minerals is more a matter of refinement than extraction. Ultimately, their work involves locating and then, as expertly as possible, separating the scarce matter they are after from all that surrounds it. As highlighted in chapters by Joshua A. Bell and Gilles Raveneau, such work always involves engaging with bulkier assemblages – landscapes, riverbeds, rock faces, geological matrices, cellphone carcasses, and so on – through processes shaped as much by responsive matter as by their own learned/shared expertise.

David Cleary's (1990) excellent account of the Amazonian gold rush of the 1980s offers glimpses of the complexity of such extractive engagements. As the *garimpeiros* with whom Cleary worked saw it, the Amazon was "a collection of micro-environments from which gold can be extracted, each one presenting characteristic technical problems and solutions" (6) necessitating a variety of skilled responses. Consider Cleary's description of how *garimpeiros* use the most important tool in their kit, the *bateia* or pan:

> Many garimpeiros will say that learning how to use a bateia takes only a few days, and it is true that the basic technique does not take long to master. But learning how to pan properly takes several months rather than days. The real skill comes when using the bateia in an area of fine grained gold. Fine grains present a problem because micro-bubbles attach themselves to the gold particles and keep them floating on the surface of the water in the bateia. This water is almost always muddy, the particles are small and difficult to see, and the inexperienced can often flush gold away. (1990: 7–8)

Refining work like this involves more than just *garimpeiros* and gold. It involves adaptive people negotiating complex networks of interaction, sloshing through muddy water, skilfully wielding *bateias*, and, ultimately, competing with micro-bubbles over the fate of practically invisible matter (Ferry & Ferry, 2017: 162–7). And that is only amidst Amazonia's secondary, alluvial, gold deposits. The engagements necessitated by gold embedded in the region's primary rock deposits are of a different, though no less complex, sort, drawing attention to how human-mineral engagements at the source are shaped not simply by specific people and minerals but also by the specificities of an ever-growing range of mineral sources that now includes discarded electronics (as discussed in Bell's chapter one on cellphone scrappers) as well as tailings pits and other contexts of circular, secondary, or urban mining.

Finding and accessing precious minerals at their sources is never assured by the techniques and technologies employed by those who seek them, as evidenced by the many ways of accounting for and attempting to mitigate inevitable uncertainties that go along with searching out needles in haystacks. As Walsh notes elsewhere, for example, some artisanal sapphire miners in Madagascar understand the sapphires they seek as having wills of their own, able to either draw in or hide from those pursuing them (Walsh, 2003). And so seekers may appeal to diviners to advise paths of underground pursuit, treat themselves with traditional medicines that will make them more attractive to what they seek, or request distant elders' or ancestors' blessing for their work. Alternatively, some (young men especially) dare to defy such conventions in pursuit of possible rewards that, for a time at least, warrant risking the anger of ancestors and the sanctioning of elders and local traditional authorities. Similarly, in diamond mining areas along the Zaire-Angola border, Filip De Boeck describes how diamonds are commonly reckoned as being "like wild animals," prone to behaving "in unpredictable and irrational ways" (1998: 186), which helps explain why they are so difficult to capture. More recently, Mette High has outlined the distinctive uncertainties and associated moral dilemmas that have come with an artisanal gold-mining rush in Uyanga, Mongolia, showing how "the power of gold" acts not only on the miners who seek it but in ways that affect and perplex miners' critics and kin, as well as the Buddhist lamas from whom they seek healing and guidance (High, 2017). In these and other cases, the pursuit of precious minerals at their sources is revealed as fundamentally unpredictable, its conduciveness to strategizing, moral dilemmas, and the elaboration of distinct ontologies indicative of the uncertainties, high stakes, and steep odds involved.[7]

What goes on at sources can also figure in reckoning the value of precious minerals from afar. The value of some precious minerals can be negatively affected by associations with conflict, child labour, unhealthy working conditions, environmental problems, and other perceived problems at the source, and, as Calvão notes with regard to diamonds later in this volume, this fact has incentivized both greater transparency and opacity in the dealings that bring diamonds to market.[8] In other cases, however, the hard, sometimes life-threatening, work involved in accessing precious minerals can add to their value, especially when highlighted as evidence of their distinctiveness, scarcity, and/or "natural" origins (see Ferry, 2005). Thus, in publications like *Colored Stone*, a glossy magazine aimed at coloured gemstone dealers and enthusiasts, and on the television channels and web pages of retailers, it isn't unusual to see pictures of pit-riddled gem fields and dirty miners alongside images of finished gemstones, the former attesting to the natural origins and great value of the latter (Walsh, 2010). Raveneau's account of Alpine crystals hunters in this volume offers another variation, presenting the risky work involved in securing particular specimens as the foundational episodes of narratives in which both people and minerals are named protagonists. In other cases, however, the work that goes into extracting precious minerals is likely to remain hidden or be obscured. The dealers and collectors with whom Ferry has done research, for example, tend to value the pristineness and naturalness of specimens extracted in Guanajuato and other locales in ways that strip them of any traces of the socio-material realities of extraction itself. The devaluation and often erasure of intermediaries between mine and collector extends this process from extraction to exchange, the process we consider next.

Exchanging Precious Minerals

In an influential contribution to the anthropology of exchange, Maurice Godelier develops a rapprochement between Marcel Mauss's theory of the gift and Marxian dialectic, based on the insight that "in order for there to be movement, exchange, there [have] to be things that are kept out of exchange, stable points around which the rest – humans, goods, services – might revolve and circulate" (Godelier, 1999: 166). Exchange, then, is not solely about the passage of goods and services from hand to hand, as a simplistically economistic reading might have it, but, more precisely, encompasses the interplay of tensions and resolutions between those things that move and those that do not (or in some cases, should not [Weiner, 1992]). Under certain arrangements, moreover, objects can combine these qualities of rootedness and movement by being exchanged only along certain channels or under very specific circumstances, such as inheritance and marriage.

Godelier writes about all sorts of objects in his analysis, but gold serves as a particularly telling instance and one that entails both movement and stasis. Using the work of Jean-Joseph Goux on gold and the gold standard, Godelier notes that gold functioned for centuries both as a medium of exchange and as a reserve, in the form of the gold standard. Furthermore, in times of crisis, its currency function would be suspended in order to maintain its ability to act as a reserve. Godelier writes: "In the very midst of a market economy, of universal currency, and generalized competition, we discover that something [gold] needs to be kept out of circulation ... for everything that can be bought and sold to begin circulating" (1999: 28). Even in contemporary finance, in which gold no longer works as a reserve currency, the presence of gold in central banks and the relationship between physical gold and other financial instruments based on the price of gold is a topic of intense and often anxious debate over the sovereignty and stability of political communities such as nation-states. In these debates as well, we see that gold continues, at the very least, to raise questions about the dialectical relationship between those things that can move and those that must stay put.

Other precious minerals also play significant and semiotically dense roles in exchange. Consider, for instance, the role of diamonds in contemporary marriage practices in the United States (and elsewhere). In her study of "the many meanings of diamonds" in the United States, Susan Falls notes the ways in which durable associations between love, marriage, and status infuse gifts of diamond engagement rings from men to women and from grandparents to grandchildren, thus enacting kin relations along two different axes (Falls, 2014).

The specific role of diamonds in the exchange relationships established in marriage is highlighted in a recent De Beers ad campaign. These ads all have a similar iconography, with white copperplate capitals against a black background, a diamond at the top, and the slogan "A Diamond Is Forever," along with the De Beers logo, at the bottom. The campaign variously invokes exchange relations, as in "Some Men Just Have That Little Extra"; "Curing Headaches Since 1888"; "Of Course There's a Return on Your Investment. We Just Can't Print It Here"; and "Exactly How Badly Do You Want to Play Golf This Weekend?" In these ads, diamonds are exchangeable for a range of things: female bodies and sexual favours, certainly, but also phalluses and male leisure. While we are not interested in arguing for a universal connection between minerals and marital/sexual exchanges, there is certainly evidence that it occurs in many different places and times. In a recent work, Wendy Doniger (2017) explores the links between sex, marriage, and jewellery (especially rings and necklaces using gold, gems, and other precious minerals) in mythology from diverse traditions.

Diamonds do not only help form conjugal and sexual transactions, they also provide assets in case those transactions break down. In Marilyn Monroe's performance of "Diamonds Are a Girl's Best Friend,"[9] she is dressed in a billowing flesh-pink dress, giving and withholding herself to a group of men dressed in black who proffer gems on black cloth. Marilyn sings of diamonds as offering women something more certain and durable than kisses: a helpful currency in the short term and, over the longer term, protection from market volatility (and the associated losses that make louses of their married admirers) and insurance against the ravages of aging (including the growing coldness of once-charmed, and no longer charming, men).

Diamonds and other precious gems also frequently act as inalienable property, incurring moral panic and distress when they are illegitimately brought into circulation. In Anthony Trollope's novel *The Eustace Diamonds*, the clever, beautiful, and unscrupulous Lizzie Eustace attempts to keep the diamond necklace that she wore while married to her husband, now deceased, in the face of the Eustace family and their lawyer, Mr Camperdown. Lizzie protests, "If a thing is a man's own he can give it away; not a house, or a farm, or a wood, or anything like that, but a thing that he can carry about with him – of course he can give it away" (Trollope & Small, 2011: 51). The diamonds in the story – and Lizzie herself – straddle the distinction between things that can and cannot be "carried about" and therefore exchanged.

More contemporary versions of the moral ambiguities of exchange occur in discussions over conflict minerals and transparency in commodity chains (Falls, 2014; Schlosser, 2013). Certification programs such as the Kimberley Process for conflict-free diamonds and the Fairmined program sponsored by the Colombia-based Alliance for Responsible Mining aim to provide incentives on ethical and market-based grounds for increased transparency and the ability to tell an ethical story about the origins of precious gems. These efforts have a new urgency and make moral-temporal claims in their own right: the emerald industry in Colombia, for instance, is seeking to establish itself as a clean, modern, and transparent economy that has moved beyond its violent past (Caraballo Acuña, 2016; but see also Brazeal, 2016b).

Precious gems, particularly those that come from the earth and are not created by humans, encapsulate movement and stasis in their very existence. Yet, much of the time, they travel far from their sources and bear no visible traces of where they come from. As Calvão has pointed out, "a diamond [like other precious minerals] actively disentangles diverging regimes of valuation ... by virtue of the liminal condition of a material extracted from nature and converted as economic value"

(2015: 205). In those instances (as in the sapphire markets studied by Walsh), when the gems from the earth and those produced in a lab are indistinguishable from each other, stones must be accompanied by stories to help us "tell the difference" (Walsh, 2010). The ways in which precious gems, such as the Millennium Sapphire described by Walsh (2010), carry their origins with them further demonstrate this dialectic between mobility and rootedness that Godelier describes as so central to exchange relations. They may also make the gems especially available for the kinds of exchanges of labour, substance, and affect generated through marriage, kinship, and political belonging. Activating such potential, however, always involves processes of valuation, to which we turn next.

The Valuation of Precious Minerals

Like other ambiguous objects,[10] minerals move along symbolic and social trajectories, getting caught up in various regimes of values as they pass through time and space. Thus, one person's cellphone may become a "treasure trove of valuable metals" for artisanal scrappers searching for gold, copper, or bronze (Bell's chapter one, this volume); a mundane blue stone may become a desirable and valuable sapphire destined to circulate in Malagasy gemstone markets and beyond (Walsh, 2004); and a mineral specimen mined in Mexico can play a central role in maintaining social relations and a sense of place, but also become a work of natural art at the Tucson Gem and Mineral Show and an object of contention in earth sciences debates questioning the power relations between Mexico and the United States (Ferry, 2013). People attend carefully to the instabilities and dynamics of minerals moving between different worlds and regimes of values. They endlessly attempt to stabilize them, at least temporarily, by evaluating their qualities, producing them as artefacts, and constantly qualifying and requalifying them (Callon, Meadel, & Rabeharisoa, 2000), not just as "precious" but also as fakes, composites, or "bricolages" in the process.

Seized in various worlds by scientists, miners, collectors, traders, lapidaries, metalworkers/founders, and other practitioners, either professional or amateur, minerals are evaluated and valued, "appraised" and "prized" (Dewey, 1939) in practice. Various sociotechnical devices take them up and contribute to their entanglement in human lives through practitioners' bodies, senses, and growing familiarity, a "becoming-minerals" (Deleuze & Guattari, 1980: 333). Which minerals' qualities do these devices highlight, work, neglect, and/or disdain? How do they produce conditions for the emergence of renewed experiences and knowledge?

How do they produce norms, classifications, and standard categories of "preciousness," despite the diversity and a "vibrant vitality" (Bennett, 2008, concerning the work of Deleuze & Guattari, 1980) of humans, minerals, and their mutual entanglement (Descola, 2013)?

Once again, diamonds give us rich and suggestive examples of these processes. Although a global commodity since antiquity (Gorelink & Gwinnett, 1998), diamonds gained the status of luxury commodity in European jewellery only during the late Middle Ages when Indians developed techniques for polishing and cutting that appealed to Venetian and later Portuguese and Dutch merchants' eyes (Hofmeester, 2012a; 2012b; 2013). From that time, the history of the diamond trade has been a long succession of monopoly claims over mining and trading operations (Hofmeester, 2012b: 20), the most famous being that of De Beers from the late nineteenth to the late twentieth century (Hofmeester, 2013: 25). As new deposits and new markets were explored through the twentieth century, however, De Beers lost its exclusive control over supply, and diamonds have become a dream accessible to (almost) everyone (Falls, 2014).

In response to shifting markets, the diamond industry struggled to come up with a universally acceptable standard for evaluating polished diamonds. Since the 1940s, the Gemological Institute of America (GIA) – founded in 1931 – has been a key leader in this process. Under the guidance of its founders, Robert M. Shipley and Richard T. Liddicoat, the GIA first developed key tools of evaluation: specialized instruments (for example, the "diamolite" and the "colorimeter"[11]), a standardized diamond grading system, and a new vocabulary and nomenclature (Copeland et al., 1960) intended "to counter the fanciful and often inconsistent terminology then being used to describe polished diamonds" (Dirlam, Shigley, & Overline, 2002: 5). Intended to ease the communication between dealers and customers, the GIA's diamond grading system was promoted (through educational programs and trade cartel practices developed in collaboration with De Beers worldwide) as the "Four Cs" (4Cs), a system of reference for diamonds that bundled colour, clarity, cut, and carat weight together.

Bundling criteria when identifying and evaluating gemstones and other minerals was not new (Sinkankas, 1986: 161). Colour and density, for example, had been used for centuries (157) in efforts at standardizing the proportions of precious metals such as gold and silver in coins, since alloys function as fiat media of exchange and commerce (see chapter six, this volume). But the emerging norms of natural sciences tend to progressively fashion, objectify, and standardize mineral properties in the process. Over recent decades, the science of gemology has developed significantly as an area of specialization for mineralogists prompted by the increasing trade of gemstones.[12] In a global trade where sophisticated treatments on natural stones as well as imitations, composites, and synthetic minerals continuously appear and spread worldwide, merchants, trade institutions, and schools have progressively incorporated tools and techniques from material sciences and chemistry in the evaluation of minerals, while the number of gem laboratories has grown to meet growing demand for measuring, characterizing, identifying, authenticating, grading, and certifying gemstones at macroscopic and microscopic scales (Fritsch & Rondeau, 2009: 147).

While the combination of the 4Cs, each with myriad gradations, can lead to the sorting of diamonds into 12,000 to 16,000 different "standardized" categories, coloured gemstones offer almost endless and (yet) non-standardized possibilities for identification and qualification due notably to their great variety of colours and internal world of inclusions. As such, "even though the appraisers use predefined guidelines to evaluate ... stones [diamonds in particular], a certain amount of subjectivity is inevitable" (Bain & Company, 2011: 38). Where some minerals, such as gold, silver, platinum, or base metals like zinc and copper, appear to be fairly homogenous and malleable substances whose properties and qualities are preserved when divided or reshaped, enabling relatively fixed prices in the global marketplace (38), diamonds and coloured gemstones are the subject of repeated evaluation and intense negotiations. The tendency towards defining and containing gemstones and other minerals in a system of measures and standardized categories, thus, is constrained by minerals' physicality as well as the limitations of human perception. Between the "naked eye" and "laboratory techniques," "professional gemologists [usually] favour simple, quick, [non-destructive] and inexpensive techniques," mainly techniques of observation (Devouard & Notari, 2009: 163–5). Instead of a distanced, objectified, or at least measurable relationship, gemstone "amateurs" (Hennion, Maisonneuve, & Gomart, 2000) seem instead to valorise attachments and seek a sensitive rapprochement with minerals' surfaces and thickness (see Vallard's chapter four, this volume). Gemologist Richard Hughes writes:

> The appraisal of precious stones is an eclectic skill, not for the timid or shameless, for such decisions involve both conscious and unconscious action. Try as we may to slice, dice and pigeon-hole elements of quality, in the end an analysis requires more than just a formula, just as fine

cooking involves not simply ingredients and a recipe. It is about reaching for factors beyond the immediate senses, and in that respect is quite like enjoyment of fine art, food and music. (2001: 1)

Hughes's thinking here unquestionably supports certain key arguments of the multibillion-dollar global natural gemstone and mineral specimen trades. To dismiss these reflections as marketing patter, however, ignores the important point that certain minerals *are* both distinctive in and of themselves and distinctively affective, to some at least. While there is no question of the role that people play in both constructing and reckoning certain minerals as precious, to assume that the pursuit of and desire for them on the part of collectors, consumers, or admirers are nothing but socially constructed outcomes of commodity fetishism risks "flatten[ing] out the passions, energies and motivations with which things" are "so fiercely invested" (Spyer, 1998: 5) by professional gemologists like Hughes, as well as by the mineralogists, visitors, and fellow anthropologists we met at the ROM. Indeed, it is often people's "passions, energies and motivations" for certain minerals – understood as being fundamentally different than other things – that most obviously mark them as precious.

In practice, there is much more to the evaluation of minerals than the categories in which industries, science, and various institutions try to contain them. Even as humans attempt to stabilize, at least for a moment, mineral properties, they are inevitably captured in a meshwork of endless uncertainties. Through authentication tests, experts experience minerals in ways that may surprise and affect them in a manner that defies established systems of qualification (Hennion & Teil, 2004: 111–38; Callon, 2002), thanks largely to the minerals themselves, whose evanescent properties can only ever be qualified, as they are experienced, in moments of contact, however mediated. Always "in-the-making" (Ingold, 2000a; 2000b; 2007) and "in the becoming" as other resources are (Richardson & Weszkalnys, 2014: 14–15, quoting Zimmermann, 1933), gemstones and minerals in general expand "existing scales for measuring and classifying a stone's potential value" (Calvão, 2013: 120). As such, certain minerals are made precious not just by their scarcity, their commodity value on global markets, or the particular, long-standing traditions of thinking in which they are reckoned so, but also by the educated attention (Ingold, 2001) and/or "educated emotion" (Hughes, 2001) paid them in "moments of valuation" (Antal, Hutter, & Stark, 2015) – including moments like the ones described at the opening of this introduction.

Organization of the Volume

When considered alongside one another, the six cases that follow suggest the diversity of human-mineral engagements at play in the world today and provide ample food for thought on how the "preciousness" of "precious minerals" might be productively conceptualized. However, some of the topics addressed and issues raised in the coming chapters bear considering apart from the whole. Accordingly, the remainder of the volume consists of two sections of clustered chapters, each introduced by a thematic overview.

Susan D. Gillespie introduces the first set of chapters by Bell, Walsh, and Raveneau with reference to what the archaeological record tells us about the long history of human engagement with minerals and how many present-day archaeologists/anthropologists like herself approach the distinctive sorts of "correspondence" (Ingold, 2013) entailed in what is, of course, an *ongoing* engagement. Bell, Walsh, and Raveneau corroborate this point in different ways in the cases that follow, all drawing attention to how moments of extraction are enabled *by* the experience and skill that people bring to the matrices in which they seek precious minerals and enabling *of* the distinct identities and/or trajectories for which specialists and the minerals they pursue are destined.

Where the chapters of the first section situate moments of extraction in ongoing processes of human-mineral entanglement, the chapters of the second section focus more on what precedes, underlies, and follows from the moments of valuation through which certain minerals are most obviously realized as precious. In introducing this second set of cases, Elizabeth Ferry draws attention first to the "constellation of meanings" concerning "the natural, the social, and the transcendental" in relation with which the minerals discussed in chapters by Vallard, Calvão, and Field make themselves and are made precious. As apparent in the entanglement of human and mineral protagonists featured in these chapters, discerning the distinctive value of precious minerals is a complex process in which both people and minerals play key roles. But Ferry also draws attention to the dangers, highlighted in Field's chapter, of assuming the inherence of preciousness, or too much of the agency of objects more generally, in how we understand the valuation of gold and other precious minerals.

To conclude the volume, we return to the topic of "preciousness" more centrally, addressing it as a quality that is best understood, appropriately enough, as multifaceted. Referring to the cases collected here, we refer specifically to what the contents of this volume reveal about preciousness, the people who reckon it, and the minerals to which it is, or might be, attributed.

NOTES

1 The highly valued "naturalness" of certain precious minerals alone could spark pages of reflection. Unquestionably a product of the socio-material relations and processes described in STS scholarship (as discussed in Ferry, 2013), this quality might also be approached with Harman's speculative realism in mind, suggestive as it is of the "inexhaustible essence" (as cited in Jensen et al., 2017: 533–4; Harman, 2013: 32) of certain things.

2 The poem was first published in England in 1968 and did not appear in the United States until almost two decades later.

3 Caillois's collection has been donated to the Muséum national d'Histoire naturelle in Paris, where part of it may be seen in the gallery of mineralogy and geology. See Caillois, 2015.

4 For Ponge, indeed, "language [is] conceived of as a way of touching" the matter at hand (Malt, 2013: 93), a medium to describe a pebble's ecology of relations and cosmogony by giving form to its inner "voice." Although humans specify matter as objects through acts of classification, to Ponge such classifications really serve as human girders, as supports, which save the poet and humanity as a whole from dizziness, allowing them to continue (Tcholakian, 1989: 94).

5 For additional reflections on how the distinctive materiality of minerals can inspire artistic creation and social analysis alike, see Leitch, 2010.

6 The conventions of gemologists and the jewellery industry that classify some minerals as precious and others as semi-precious or something else are not fixed. As the gemologist Richard Wise notes, "the whole idea of preciousness is fluid" in the gemstone industry. "In the world of gemstones, if it is rare and beautiful, and if demand is strong, it is precious" (Wise, 2006: 11).

7 Such ambiguity, of course, also features away from sources as described, for example, in Brazeal's (2017) account of how Jains involved in the global emerald trade pursue both profit and salvation (through austerity) through their work.

8 For an intriguing counter-example to commonly invoked narratives on conflict minerals, see Brazeal's (2016b) account of the complex effects of war and peace on Colombia's emerald trade.

9 The lyrics of this song were written by Wendy Doniger's uncle! (Doniger, personal communication).

10 Shells, for example, exist at the frontier between life and death, mundaneness and preciousness. See Faugère & Senépart, 2012.
11 See Shipley & Liddicoat, 1941 and Liddicoat, 1981. By the 1930s, the GIA had other patented instruments to examine the interior of gems, such as triple aplanatic lens loupes and gemological microscopes.
12 In 2009, the global gemstone trade was valued at US$20 to 25 billion, about 85 per cent of which is accounted for solely by diamonds (Fritsch & Rondeau, 2009: 147).

REFERENCES

Antal, A.B., Hutter, M., & Stark, D. (2015). *Moments of valuation: Exploring sites of dissonance*. Oxford: Oxford University Press.

Appadurai, A. (Ed.). (1986). *The social life of things*. Cambridge: Cambridge University Press.

Appadurai, A. (2006). The thing itself. *Public Culture, 18*(1), 15–21. https://doi.org/10.1215/08992363-18-1-15

Bain & Company. (2011). *The global diamond industry: Lifting the veil of mystery*. Antwerp: AWDC.

Balfet, H. (Ed.). (1991). *Observer l'action technique: Des chaînes opératoires pour quoi faire?* Paris: CNRS.

Bennett, J. (2008). Matérialismes métalliques. *Rue Descartes, 59*(1), 57–66. https://doi.org/10.3917/rdes.059.0057

Boivin, N., & Owoc, M.A. (2004). *Soil, stones and symbols: Cultural perceptions of the mineral world*. London: UCL.

Brazeal, B. (Dir.). (2016a). *Illusions in stone: The global story of the emerald trade*. Produced by the Advanced Laboratory for Visual Anthropology. California State University, Chico. Chico: ALVA Productions, 58 min. Retrieved from www.csuchico.edu/alva/projects/2016/illusions-in-stone.shtml

Brazeal, B. (2016b). Nostalgia for war and the paradox of peace in the Colombian emerald trade. *The Extractive Industries and Society, 3*(2), 340–9. https://doi.org/10.1016/j.exis.2015.04.006

Brazeal, B. (2017). Austerity, luxury and uncertainty in the Indian emerald trade. *Journal of Material Culture, 22*(4), 437–52. https://doi.org/10.1177/1359183517715809

Caillois, R. (1985). *The writing of stones*. B. Bray (Trans.). Charlottesville: University Press of Virginia. (Original work published in 1970: Callois, R. *L'écriture des pierres*. Genève: Albert Skiar)

Caillois, R. (2015). *La lecture des pierres*. Paris: Éditions Xavier Barral, Muséum national d'Histoire naturelle.

Callon, M. (2002). Pour en finir avec les incertitudes? *Sociologie du travail, 44,* 261–7.

Callon, M., Meadel, C., & Rabeharisoa, V. (2000). L'économie des qualities. *Politix, 13*(52), 211–39. https://doi.org/10.3406/polix.2000.1126

Calvão, F. (2013). The transporter, the agitator and the kamanguista: Qualia and the in/visible materiality of diamonds. *Anthropological Theory, 13*(1–2), 119–36. https://doi.org/10.1177/1463499613483404

Calvão, F. (2015). Diamonds, machines and colours: Moving materials in ritual exchange. In A. Drazin & S. Küchler (Eds.), *The social life of materials: Studies in materials and society* (pp.193–208). London: Bloomsbury.

Caraballo Acuña, V. (2016). Comerciar sin afiebrarse. Cronotopos, qualias y oposiciones en la formalización de la explotación y el comercio de esmeraldas en Colombia. Unpublished master's thesis, Colegio de Michoacán, Centro de Estudios Antropológicos.

Casanova, M. (2013). *Le lapis-lazuli dans l'Orient ancient: Production et circulation du néolithique au IIe millénaire av. J.C.* Paris: CTHS.

Cauvin, M.C., Gourgaud, A., Gratuze, B., Arnaud, N., Poupeau, G., Poidevin, J.L., & Chataigner, C. (Eds.). (1998). *L'obsidienne au Proche et Moyen Orient du volcan à l'outil.* BAR International Series, 738. Oxford: Archaeopress.

Cleary, D. (1990). *Anatomy of the Amazon gold rush.* New York: Springer.

Cohen, J.J. (2015). *Stone: An ecology of the inhuman.* Minneapolis: University of Minnesota Press.

Copeland, L.L., Liddicoat, R.T., Benson, L.B., Martin, J.G.M., & Crowningshield, G.R. (1960). *The diamond dictionary.* Los Angeles, CA: Gemological Institute of America.

De Boeck, F. (1998). Domesticating diamonds and dollars: Identity, expenditure and sharing in southwestern Zaire (1984–1997). *Development and Change, 29*(4), 777–810. https://doi.org/10.1111/1467-7660.00099

Deleuze, G., & Guattari, F. (1980). *Mille plateau.* Paris: Edition de Minuit.

Descola, P. (2013). *Beyond nature and culture.* J. Lloyd (Trans.). Chicago: University of Chicago Press. (Original work published 2005: Descola, P. *Par-delà nature et culture.* Paris: Gallimard)

Devouard, B., & Notari, F. (2009). The identification of faceted gemstones: From the naked eye to laboratory techniques. *Elements, 5*(3), 163–8. https://doi.org/10.2113/gselements.5.3.163

Dewey, J. (1939). *Theory of valuation: Vol. II, no. 4. International encyclopaedia of unified science.* Chicago: University of Chicago Press.

Dirlam, D.M., Shigley, J.E., & Overline, S.T. (2002). The ultimate gemologist: A tribute to Richard T. Liddicoat. *Gems & Gemology, 38*(1), 2–13. Retrieved from https://www.gia.edu/gems-gemology/spring-2002-tribute-richard-liddicoat-dirlam

Doniger, W. (2017). *The ring of truth and other myths of sex and jewelry*. Oxford: Oxford University Press.

Falls, S. (2014). *Clarity, cut, and culture: The many meanings of diamonds*. New York: NYU Press.

Faugère, E., & Senépart, I. (Eds.). (2012). *Itineraries of shells*, vol. 59 of *Techniques & Culture*. https://doi.org/10.4000/tc.6341

Ferry, E.E. (2005). Geologies of power: Value transformations of mineral specimens from Guanajuato, Mexico. *American Ethnologist*, 32(2), 420–36. https://doi.org/10.1525/ae.2005.32.3.420

Ferry, E.E. (2013). *Minerals, collecting, and value across the U.S.-Mexican border*. Bloomington: University of Indiana Press.

Ferry, E.E. (2015, 14 July). The life and times of minerals. (Blog post). Retrieved from http://blog.castac.org/2015/07/the-life-and-times-of-minerals/

Ferry, E.E., & Ferry, S.E. (2017). La batea. Brooklyn, NY: Red Hook Editions.

Fritsch, E., & Rondeau, B. (2009). Gemology: The developing science of gems. *Elements: An International Magazine of Mineralogy, Geochemistry and Petrology*, 5(3), 147–52. https://doi.org/10.2113/gselements.5.3.147

Godelier, M. (1999). *The enigma of the gift*. Chicago: University of Chicago Press.

Golub, A. (2014). *Leviathans at the gold mine: Creating indigenous and corporate actors in Papua New Guinea*. Durham, NC: Duke University Press.

Gorelink, L., & Gwinnett, A.J. (1998). Diamonds from India to Rome and beyond. *American Journal of Archaeology*, 92(4), 547–52. https://doi.org/10.2307/505249

Hacking, I. (2002). *Representing and intervening: Introduction topics in the philosophy of the natural sciences*. Cambridge: Cambridge University Press.

Haraway, D. (2003). *The companion species manifesto: Dogs, people, and significant otherness*. Chicago: University of Chicago Press.

Harman, G. (2013). *Bells and whistles: More speculative realism*. Winchester, UK: Zero Books.

Helmreich, S. (2009). *Alien ocean: Anthropological voyages in microbial seas*. Berkeley: University of California Press.

Hennion, A., Maisonneuve, S., & Gomart, E. (2000). *Figures de l'amateur: Formes, objets, pratiques de l'amour de la musique aujourd'hui*. Paris: La documentation française.

Hennion, A., & Teil, G. (2004). Le goût du vin: Pour une sociologie de l'attention. In V. Nahoum-Grappe & O. Vincent (Eds.), *Le goût des belles choses*. Paris: MSH.

High, M. (2017). *Fear and fortune: Spirit worlds and emerging economies in the Mongolian gold rush*. Ithaca, NY: Cornell University Press.

Hofmeester, K. (2012a). Les diamants, de la mine à la bague: Pour une histoire globale du travail au moyen d'un article de luxe. *Le Mouvement Social*, 241(4), 85–108. https://doi.org/10.3917/lms.241.0085

Hofmeester, K. (2012b). Working for diamonds from the 16th to the 20th century. In M. Van der Linden & L. Lucassen (Eds.), *Working on labor: Essays in Honor of Jan Lucassen* (pp.19–46). Boston, MA: Brill.

Hofmeester, K. (2013). Shifting trajectories of diamond processing: From India to Europe and back, from the fifteenth century to the twentieth. *Journal of Global History*, *8*(1), 25–49. https://doi.org/10.1017/s174002281300003x

Hughes, R. (2001). Passion fruit: A lover's guide to sapphire. *The Guide*, *21*(2), 3–15. Retrieved from http://www.palagems.com/sapphire-connoisseurship

Ingold, T. (2000a). Making culture and weaving the world. In P.M. Graves-Brown (Ed.), *Matter, materiality and modern culture* (pp. 50–71). London: Routledge.

Ingold, T. (2000b). *The perception of the environment: Essays on livelihood, dwelling and skill*. London: Routledge.

Ingold, T. (2001). From the transmissions of representations to the education of attention. In H. Whitehouse (Ed.), *The debated mind: Evolutionary psychology versus ethnography* (pp.113–53). Oxford: Berg.

Ingold T. (2007). Materials against materiality. *Archaeological Dialogues*, *14*(1), 1–16. https://doi.org/10.1017/s1380203807002127

Ingold, T. (2013). *Making: Anthropology, archaeology, art and architecture*. London: Routledge.

Jensen, C.B., Ballestero, A., de la Cadena, M., Fisch, M., & Ishii, M. (2017). New ontologies? Reflections on some recent "turns" in STS, anthropology and philosophy. *Social Anthropology/Anthropologie sociale*, *25*(4), 525–45. https://doi.org/10.1111/1469-8676.12449

Jorgensen, D. (1997). Who and what is a landowner? Mythology and marking the ground in a Papua New Guinea mining project. *Anthropological Forum*, *7*(4), 599–627. https://doi.org/10.1080/00664677.1997.9967476

Kirsch, S. (2014). *Mining capitalism: The relationship between corporations and their critics*. Oakland: University of California Press.

Kohn, E. (2013). *How forests think: Toward an anthropology beyond the human*. Berkeley: University of California Press.

Kopytoff, I. (1986). The cultural biography of things: Commoditization as process. In A. Appadurai (Ed.), *The social life of things: Commodities in cultural perspective* (pp. 64–92). Cambridge: Cambridge University Press. doi:10.1017/CBO9780511819582.004

Leitch, A. (2010). Materiality of marble: Explorations of the artistic life of stone. *Thesis Eleven*, *103*(1), 65–77. https://doi.org/10.1177/0725513610381375

Lemonnier, P. (1976). La description des chaînes opératoires: Contribution à l'analyse des systèmes techniques. *Techniques et Culture* (bulletin), 1, 100–51.

Leroi-Gourhan, A. (1965). *Le geste et la parol: tome 2. La mémoire et les rythmes*. Paris: Albin Michel.

Liddicoat, R.T. (1981). *Handbook of gem identification*. Santa Monica, CA: Gemological Institute of America. (Original work published 1947)

Malt J. (2013). Leaving traces: Surface contact in Ponge, Penone and Alÿs. *Journal of Verbal/Visual Enquiry, 29*(1), 92–104. https://doi.org/10.1080/02666286.2012.746265

Nash, J.C. (1993). *We eat the mines and the mines eat us: Dependency and exploitation in Bolivian tin mines*. New York: Columbia University Press.

Niedecker, L. (2013). *Lake Superior*. Seattle, WA: Wave Books.

Palsson, G., & Swanson, H.A. (2016). Down to earth: Geosocialities and geopolitics. *Environmental Humanities, 8*(2), 149–71. https://doi.org/10.1215/22011919-3664202

Peluso, N.L. (2018). Entangled territories in small-scale gold mining frontiers: Labor practices, property, and secrets in Indonesian gold country. *World Development, 101*, 400–16. https://doi.org/10.1016/j.worlddev.2016.11.003

Pentcheva, B.V. (2006). The performative icon. *The Art Bulletin, 88*(4), 631–55. https://doi.org/10.1080/00043079.2006.10786312

Peterson, K. (2013). Fossil necklace. Retrieved from http://katiepaterson.org/portfolio/fossil-necklace/

Ponge, F. (1948). Notes premières de l'homme. In *Proêmes*. Paris: NRF.

Ponge, F. (1971). *Méthodes*. Paris: Gallimard. (Original work published 1961)

Ponge, F. (1974). The pebble. B. Archer (Trans.). In *The voice of things* (pp. 69–77). New York: McGraw Hill. (Original work published 1942: Ponge, F. *Le parti pris des choses*. Paris: Gallimard)

Povinelli, E.A. (2016). *Geontologies: A requiem to late liberalism*. Durham, NC: Duke University Press.

Raffles, H. (2012). Twenty-five years is a long time. *Cultural anthropology, 27*(3), 526–34. https://doi.org/10.1111/j.1548-1360.2012.01158.x

Rajak, D. (2011). *In good company: An anatomy of corporate social responsibility*. Stanford, CA: Stanford University Press.

Revolon, S. (2012). L'Éclat des ombres: Contraste, iridescence et présence des morts aux îles Salomon. *Techniques & Culture, 58*: 252–63. https://doi.org/10.4000/tc.6299

Richardson, T., & Weszkalnys, G. (2014). Introduction: Resources materialities. *Anthropological Quarterly, 87*(1), 5–30. https://doi.org/10.1353/anq.2014.0007

Roberts, B.W., & Thornton C. (2014). *Archeometallurgy in global perspective: Methods and syntheses*. New York: Springer.

Rolston, J.S. (2013). The politics of pits and the materiality of mine labor: Making natural resources in the American West. *American Anthropologist, 115*(4), 582–94. https://doi.org/10.1111/aman.12050

ROM. (2018). Michael Lee-Chin crystal. Retrieved from https://www.rom.on.ca/en/about-us/rom/michael-lee-chin-crystal

Salas Carreño, G. (2017). Mining and the living materiality of mountains in Andean societies. *Journal of Material Culture*, *22*(2), 133–50. https://doi.org/10.1177/1359183516679439

Saunders, N. (1999). Biographies of brilliance: Pearls, transformations of matter and being, c. AD 1492. *World Archaeology*, *31*(2): 243–57. https://doi.org/10.1080/00438243.1999.9980444

Schlosser, K. (2013). Regimes of ethical value? Landscape, race and representation in the Canadian diamond industry. *Antipode*, *45*(1), 161–79. https://doi.org/10.1111/j.1467-8330.2012.00996.x

Shipley R.M., & Liddicoat R.T. (1941). A solution to diamond color grading problems. *Gems & Gemology*, *3*(11), 162–7. Retrieved from https://www.gia.edu/doc/A-Solution-to-Diamond-Color-Grading-Problems.pdf

Simondon, G. (1958). *Du mode d'existence des objets techniques*, Paris: Ed. Aubier-Montaigne.

Sinkankas, J. (1986). Contributions to a history of gemology: Specific gravity – origins and development of the hydrostatic method. *Gems & Gemology*, *22*(3), 156–65. https://doi.org/10.5741/gems.22.3.156

Smith, J.H. (2011). Tantalus in the digital age: Coltan ore, temporal dispossession, and "movement" in the Eastern Democratic Republic of the Congo. *American Ethnologist*, *38*(1), 17–35. https://doi.org/10.1111/j.1548-1425.2010.01289.x

Spyer, P. (1998). Introduction. In P. Spyer (Ed.), *Border fetishisms: Material objects in unstable spaces* (pp. 1–11). New York: Routledge.

Tcholakian, M.T. (1989). La pierre dans la poésie de Ponge. *Etudes françaises*, *25*(1), 89–113. https://doi.org/10.7202/035775ar

Tilley, C. (2004). *The materiality of stone: Explorations in landscape phenomenology*. Oxford: Berg Publishers.

Trébuchet-Breitwiller, A.-S. (2011). *Le travail du précieux: Une anthropologie économique des produits de luxe à travers l'exemple du parfum et du vin*. Thèse de doctorat en économie et finances. École Nationale Supérieure des Mines de Paris. Retrieved from https://pastel.archives-ouvertes.fr/pastel-00713690

Trollope, A., & Small, H. (2011). *The Eustace diamonds*. Oxford: Oxford University Press.

Tsing, A.L. (2000). Inside the economy of appearances. *Public Culture*, *12*(1), 115–44. https://doi.org/10.1215/08992363-12-1-115

Tsing, A.L. (2015) *The mushroom at the end of the world: On the possibility of life in capitalist ruins*. Princeton, NJ: Princeton University Press.

Walsh, A. (2003). "Hot money" and daring consumption in a Northern Malagasy mining town. *American Ethnologist*, *30*(2), 290–305. https://doi.org/10.1525/ae.2003.30.2.290

Walsh, A. (2004). In the wake of things: Speculating in and about sapphires in Northern Madagascar. *American Anthropologist, 106*(2), 225–37. https://doi.org/10.1525/aa.2004.106.2.225

Walsh, A. (2010). The commodification of fetishes: Telling the difference between natural and synthetic sapphires. *American Ethnologist, 37*(1), 98–114. https://doi.org/10.1111/j.1548-1425.2010.01244.x

Weiner, A.B. (1992). *Inalienable possessions: The paradox of keeping-while giving.* Berkeley: University of California Press.

Wise, R.W. (2006). Secrets of the gem trade: The connoisseur's guide to precious gemstones. Lennox, MA: Brunswick House Press.

Wortham, S. (1996). Sovereign counterfeits: The trial of the pyx. *Renaissance Quarterly, 49*(2): 334–59. https://doi.org/10.2307/2863161

Yourcenar, M. (1985). Introduction. In R. Caillois, *The writing of stones.* B. Bray (Trans.). Charlottesville: University Press of Virginia.

Zalasiewicz, J. (2012). *The planet in a pebble: A journey into earth's deep history.* Oxford: Oxford University Press.

Zimmermann, E. (1933). World resources and industries: A functional appraisal of the availability of agricultural and industrial resources. New York: Harper & Row.

PART ONE

Engaging Mineral Sources

Introduction to Part One
Scrappers, Miners, and Hunters

SUSAN D. GILLESPIE

This first set of chapters focuses on the work of artisanal miners to extract precious minerals from their matrixes, exposing the brute physicality and social implications of miner-mineral entanglements. Humans have been so entangled since at least the Neolithic period, when digging into the earth for materials became more commonplace, creating trajectories of human-thing interdependencies (Hodder, 2012). That new "economy of substances" (Thomas, 1999: 74) involved the circulation of materials in an emergent system of signification that gave rise to new identities, and the same still occurs today.

Neolithic peoples became entangled primarily with useful, mundane, and oftentimes ubiquitous earthy substances such as clay, chalk, and stone (Boivin, 2004; Conneller, 2011; Hodder, 2012; Thomas, 1999). Yet, during this same era, copper, gold, and jade – precious minerals – were first exploited for their distinctive properties, conjuring new categories of persons and social relationships (Renfrew, 2001). From those early beginnings, minerals and gems were set "in-motion" (Appadurai, 1986: 5), becoming central to shifting assemblages engaging human and non-human actants (Latour, 2005). Once polished, primed, or primped, these materials tend to retain their irresistible qualities. They may move with less "friction" and they age well – they "perdure," not in a state of stasis but as a means to gain new leases on life (Ingold, 2013: 104). Precious minerals are not limited to mere symbolic functions but bundle specific physical properties with indexical associations that play their own roles in social interactions (Renfrew, 2001: 131; Thomas, 1999: 73). As revealed throughout this volume's studies, the value of precious minerals is neither inherent nor strictly economic, but is constantly gauged and reassessed within multiple interconnected socio-material assemblages.

These three chapters – on cellphone scrapping in the United States (Joshua A. Bell), sapphire mining in Madagascar (Andrew Walsh), and crystal hunting in the Alps (Gilles Raveneau) – detail how extraction is an act of disassembly that requires assembling local and historically specific actors: people, materials, tools, and places. Operating as individuals or as members of a loose community, the miners and the minerals they seek come into a relation with one another, one which Ingold (2013: 70) characterizes as a "correspondence" foregrounded in the anticipated use of the mineral as "precious." This physical work engages certain bodily motricities as well as other acquired experiential skills (Warnier, 2001), such as knowing where to look and whom to look out for.

The minerals themselves have their own agencies and can act in obstreperous ways, resisting attempts to encounter, dislodge, and carry them away. Their rarity or scarcity requires patience and forbearance, as well as a bit of luck and possibly supernatural intervention, to accomplish the goal. Landscape features, rocky matrixes, and the dispersal of minerals in concentrated veins, Alpine "ovens," or cellphones are themselves principal actants that impact the decisions, actions, and success of the extractors. The work is fraught with uncertainties, risks, and the threat of bodily peril, even death – an affective quality that hangs over the entire enterprise. Risk intersects with the marginalized positioning of some miners, although it may also play a role in elevating their status.

Once obtained, the minerals and the miners are still "in-motion," forming new assemblages necessary to assess or realize the mineral's economic value, even as it valuates the miners themselves. As Walsh indicates, this process is another kind of "correspondence," now with fellow practitioners and markets as well as collectors and publics. This assemblage can be characterized by competition, rivalry, and conflict, as well as cooperation and collaboration, and is coloured by unpredictability and risk, like the extractive activities themselves. These two assemblages are interdependent: without the markets and institutionalized standards to monetize the minerals, there would be no mining; but the scarcity, risk, and – very often – the stories of their extraction combine with the minerals' physical qualities in their eventual valuation.

As a consequence, the entanglement with minerals mutually constitutes the miners and the networks of their extraction and trade, not to mention the objects themselves. A major theme of these chapters is the intra-subjectivation processes that make both the miners and the minerals by their participation in these assemblages. Distinctive subject positions emerge from the relational networks of people and materials (Thomas, 1999: 72). Techniques of a skilled body are configured as "techniques of the self" (Warnier, 2001), although these identities may be recognized primarily only within the community of practitioners. If the artisanal extraction of precious minerals is a mode of self-making, it is not strictly autopoietic. It requires the intervention, engagement, and sharing of ontological properties with other actors in dynamic assemblages, some of them fleeting, others institutionalized.

Scrappers

In Bell's analysis (chapter one), contemporary American artisanal e-waste scrappers who salvage gold and other metals from cellphones "remake themselves" by becoming and especially by *being seen* as experts. Unlike other miners, scrappers are usually autonomous individuals. They operate at some distance from one another, given that they extract metals from portable objects with no geographic restrictions. Cellphones are usually black boxes, not meant to be opened – and, indeed, scrappers may employ brute force to break them apart. Bodily skill and technical (or alchemical) knowledge are required to recognize and retrieve trace amounts of valuable metals hidden inside. These materials are rather comparable in their small size and concentrated value to gemstones, although, as metals destined for recycling, they are more fungible. As is the case with other artisanal miners, scrapping is more than just a hobby, for many truly need the income. Ironically, some scrappers earn more from disseminating their knowledge of scrapping than from selling the precious metals.

Acknowledging that they are engaged in salvage work considered secondary, dirty, and downright dangerous due to toxic chemicals needed to extract the gold, scrappers seek to acquire a certain legitimacy for this form of mining. Increasingly, their expertise is gauged in terms of the performances of the body as performances of the self (Goffman, 1956). Their extractive performances and expert advice are filmed and uploaded as instructional videos to YouTube, witnessed asynchronously by a social media audience Bell calls "scrapper publics." By following several digital micro-celebrity scrappers – male and female – using their videos and statements as well as comments from their publics, Bell perceptively penetrates the world of online scrapping.

Many scrappers deploy monikers that disguise, exaggerate, and even fabricate their personae as scrapper experts in their videos. The videos become the foci of a virtual assemblage of social media platforms and the dispersed followers who watch the videos and record their witnessing in posts on online forums, websites, and Facebook. The videos highlight how the scrappers' extractive actions and expertise are performative, embodied, and linguistically enacted, and must be routinely demonstrated and reaffirmed by the practitioners. In addition, using the networking capabilities of YouTube, some scrappers link their videos to one another, in part to drive more traffic to their own performances. This competition for prestige parallels, in some ways, that of the communities of Alpine crystal hunters in Raveneau's chapter.

Stories, including autobiographies, are essential to the valuation of the miner (but not the mineral – a contrast with the chapters by Walsh and Raveneau). However, in this case they emerge in dynamic dialogues between the scrappers and their publics. The videos and websites, with their comments and responses, are "cultural texts," which, as Bell says, "reveal attitudes and tensions in the work at hand and show how individuals are aiming to position themselves as experts and as part of a wider network of practitioners."

Through these texts, scrappers offer insights into the dynamic interconnections occurring on the margins of the American economy in the revaluation of accumulating discarded electronic debris. They are engaged in self-making in a peculiarly American manner, calling on the tropes of hope, desire, and freedom. YouTube scrappers explicitly invoke the American dream of doing better, even as it requires dismantling icons of American progress and business success. Scrapping is simultaneously an act of revolutionary iconoclasm and of making money in the most idyllic capitalist way: out of something that is mostly free. For scrappers, transforming free things into money is a way to assert their ability to achieve freedom.

Miners

In his ethnographic study of sapphire miners in Madagascar (chapter two), Walsh usefully questions the "artisanal" aspect of what are more usually lumped together as "artisanal and small-scale mining" (ASM) producers. The "small-scale" characterization makes sense when contrasted with industrial-size, government-sanctioned mining operations, but what does it mean to be an "artisanal" gemstone miner? The obvious but until now elusive answer reached by Walsh is to compare them with other artisans, which means to focus on their engagements with the materials they manipulate, as well as on their situatedness in local and global relationships that facilitate and valuate their extractive production.

Walsh understands the artisanal miners of Madagascar as skilled practitioners rather than exploited subalterns (compare to Bell's scrappers). Employing assimilated body techniques, these "autotelic" – that is, self-propelled – actors create distinct niches of opportunity at the fringes of the more industrial-scale and legitimate mining enterprises. Where the similarity with often romanticized notions of artisanal craftspeople (think cheesemakers) ends is the recognition that other kinds of knowledge and skills are required to outdo rivals and to navigate complex market and governmental forces on a global scale. Walsh aims to approach artisanal production in a way that attends to how artisans engage simultaneously in distinct sorts of "correspondences" (following Ingold) with the materials of their work and the landscapes in which they occur, as well as in distinctive sorts of competition, collaboration, and conflicts with other people. As he emphasizes, the miners are shaped both by the flow of the materials with which they correspond in their work and by the markets for their products.

Thus, Walsh details the properties and circumstances of the coloured gemstones themselves. He begins, unconventionally, with the gemological report on corundum (the mineral of both sapphires and rubies) for the Ankarana region, which has been the focus of his ethnographic research. The mineral's material properties and the features of the landscapes in which it occurs are salient factors in artisanal mining and the marketing of these gemstones. The sheer fact that gemstones could be found with shovels in alluvial deposits facilitated their extraction – the tools required were simple enough – but this assemblage is dynamic and shifting, requiring constant adjustments in the correspondences between miners and the ever more elusive gemstones.

Since the turn of the twenty-first century, artisanal sapphire mining has become more precarious, unpredictable, and dangerous as the gemstones seem to increasingly resist being found. Miners must work in teams, often supported by a patron, forming new or more intense and prolonged social assemblages necessary for digging into caves and excavating deep pits in the unstable alluvium. Months may go by without a find. Furthermore, because sapphires are inherently distinct, varying greatly in size, colour, visual appearance, crystal morphology, and so forth, there is far greater speculation in their valuation. The markets for sapphires are uncertain as demand and prices fluctuate; thus, the work of gem traders is equally precarious.

Nevertheless, the gems retain certain "potentials": the concentrated value of their small size, their hardness, and (once primped) the endurance of their shape and polish – potentials that are constantly weighed against the uncertainty and risk of their extraction and trade. Although synthetic sapphires can be produced in a laboratory and are almost indistinguishable from those taken from the ground, gems with "natural" origins remain most prized. That aura, an indexical link to the earth and to the work of miners, is essential to their valuation. Thus, the materiality of sapphires and their landscapes contribute to miner-sapphire entanglements in complex assemblages impacting the global gemstone market.

Hunters

Seekers of natural crystals in the Alps refer to themselves as hunters, as if stalking a wary prey. Although quartzite and other crystalline minerals took eons to form and are difficult to pry from their stony matrix, Raveneau (chapter three) treats them as things-in-motion (following Appadurai), whose individual trajectories or "social lives" should be followed. Their value and ontological status can shift dramatically once they are dislodged from the mountains and enter – or not – the market exchange. Raveneau emphasizes how a symbolic production is added to this economic production, one that can make or undo the status and reputation of crystal hunters as well as that of the crystals, as the latter move between hunters and collectors or come to rest, at least temporarily, in museums.

Crystal hunting is a risky business. Many hunters work in small teams, in competition with one another over territory. The task requires a great deal of skill and experience for success, not to mention the sheer stamina of high altitude climbing over rough terrain for long periods. Formed in veins of quartz, high quality crystals typically can be located precisely in the areas of crumbly walls and "rotten" rocks avoided by hikers for the dangers they present. These places are where cracks appear in the earth's surface that allow access into its depths, especially natural cavities ("ovens") where the best crystals are hiding.

The hunters are thus at the mercy of the natural environment and weather conditions, incurring injuries and sometimes fatal falls. As Raveneau explains, they are egged on by a treasure quest mentality. Stories circulate of a "virgin 'oven,' the miraculous crack full of 'mature' crystals that have been waiting ... for millions of years, which [the hunters] will be able to take down intact and pristine." The gendered and treasure-seeking tropes of crystal hunting are not unlike that of (mostly) male scrappers busting into and taking the pristine-looking gold from the "motherboard" of a cellphone. Hunters bet that their lives will not be taken in the acquisition of crystals; however, sacrificed lives increase the overall value of crystals as the social lives of hunters and minerals become entangled.

Unlike diamonds and recycled gold, dislodged crystals are singular objects, no two exactly identical, and some are quite extraordinary. The exceptional ones are given individuating proper names, usually that of their finder or someone else involved in their discovery or pedigree of ownership; the story of their history is a component of their singularity. Raveneau hypothesizes that crystals constitute a "principle of identity" that can be linked to anthropological theories of the Maussian inalienable gift and prestige goods.

Although most hunters are hobbyists, they will sell or trade their finds, although they may also keep them. Nevertheless, what hunters seek is prestige and fame within the "community field" or assemblages of fellow crystal hunters and collectors. Monetary gains from selling crystals contribute to hunters' fame as recognition of their skill. As Raveneau notes, the "social value" that differentiates the more esteemed hunters is tied up with their manhood, honour, power, "and other relevant concepts that crystallize in hunters' relationships with one another." In turn, the value of the crystal derives not just from its aesthetic qualities or uniqueness, but also from the history of its discovery. The personhood of the discoverer is inalienably indexed in the crystal; thus, its exchange value cannot encompass its total value.

Conclusion

These three authors, in very different studies, acknowledge implicitly or explicitly that there are no strictly "social" relations (Gosden, 1999: 120). Non-human agents are enmeshed in all human projects – here, the materials, landscapes, made objects, and texts (and other media) that act as framing or focusing devices. In following miners and minerals as things-in-motion, Bell, Walsh, and Raveneau demonstrate how situated interactions with precious minerals "make" the artisanal miners. Even these small-scale or hobbyist extractors are entangled with the properties and circumstances of the minerals themselves and the external markets and institutional structures within which their products are exchanged and evaluated.

REFERENCES

Appadurai, A. (1986). Introduction: Commodities and the politics of value. In A. Appadurai (Ed.), *The social life of things: Commodities in cultural perspective* (pp. 3–63). Cambridge: Cambridge University Press.

Boivin, N. (2004) From veneration to exploitation: Human engagement with the mineral world. In N. Boivin & M. Owoc (Eds.), *Soils, stones and symbols: Cultural perceptions of the mineral world* (pp. 1–30). London: University College London Press.

Conneller, C. (2011). *An archaeology of materials: Substantial transformations in early prehistoric Europe.* London: Routledge.

Goffman, E. (1956). *The presentation of the self in everyday life*. New York: Random House.

Gosden, C. (1999). *Anthropology and archaeology: A changing relationship*. London: Routledge.

Hodder, I. (2012). *Entangled: An archaeology of the relationships between humans and things*. Malden, MA: John Wiley & Sons.

Ingold, T. (2013). *Making: Anthropology, archaeology, art and architecture*. London: Routledge.

Latour, B. (2005). *Reassembling the social: An introduction to actor-network-theory*. Oxford: Oxford University Press.

Renfrew, C. (2001). Symbol before concept: Material engagement and the early development of society. In I. Hodder (Ed.), *Archaeological theory today* (pp. 122–40). Malden, MA: Blackwell.

Thomas, J. (1999). An economy of substances in earlier Neolithic Britain. In J.E. Robb (Ed.), *Material symbols: Culture and economy in prehistory* (pp. 70–89). Occasional papers, vol. 26. Center for Archaeological Investigations. Carbondale: Southern Illinois University.

Warnier, J.P. (2001). A praxeological approach to subjectivation in a material world. *Journal of Material Culture*, 6(1), 5–24. https://doi.org/10.1177/135918350100600101

1 "Check Out That Gold-Plated Board!" Scrapping Cellphones and Electronics in North America

JOSHUA A. BELL

"Here we have a man whose job it is to gather the day's refuse in the capital. Everything that the big city has thrown away, everything it has lost, everything it has scorned, everything it has crushed underfoot he catalogues and collects. He collates the annals of intemperance, the capharnaum of waste. He sorts things out and selects judiciously: he collects like a miser guarding a treasure, refuse which will assume the shape of useful or gratifying objects between the jaws of the goddess of Industry." This description is one extended metaphor for the poetic method, as Baudelaire practiced it. Ragpicker and poet: both are concerned with refuse, and both go about their solitary business while other citizens are sleeping.

Walter Benjamin,
"The Paris of the Second Empire in Baudelaire" (2003: 48)

Google the phrase "scrapping cellphone," and a 3 minute, 32 second video entitled "Scrapper Girl scraps a cell phone for gold and talks about Top Dollar Mobile" appears.[1] Originally published on 7 September 2012, as of 15 September 2016 the YouTube video had received 637,048 views with 515 comments.[2] The video begins with a montage of a young white blonde woman making various faces as the names of different metals – GOLD, SILVER, BRASS, COPPER, ALUMINUM, STEEL, BRONZE, IRON – are displayed until her name is proclaimed: "SCRAPPER GIRL." Shifting to the inside of a workshop, the woman looks at the camera and says:

Hey everybody, Scrapper Girl here. Today, I'm gonna show you guys how to scrap a cellphone ... We're gonna do this because inside we want *to get the gold-plated board and various other metals that might be inside* ... We'll start by taking the back off, just gotta press this button here. Then make sure you pop the battery out – don't want to be dealing with that. Inside you'll find some tiny little screws that you gotta get out. There's one in here and there's another. Come on ... Alright, I think that should be good. (emphasis added)

Opening the phone's back with a screwdriver, Scrapper Girl continues:

We just have to pop this back off. They can be a little tricky, but you just gotta pull [struggles to get it off – but it opens] ... Alright ... then there should be some tiny screws in here that you're gonna have to take out as well. Four of them, I believe. Oops! [She drops a screw.] All the cellphones are going to be a bit different, but you just got to make sure you take them all out and there we go!

The cellphone open, Scrapper Girl does the grand reveal, holding the phone's motherboard to the camera:

Check out that gold-plated board! That's the good stuff – that is what we are looking for. Before we start doing something with that, we have to take this screen off ... and save it, as there are some precious metals in there that you can get ... In this part of the phone there's gold plating inside the speaker, so if you wanna go ahead and take it apart that ... works too. Some of you guys might want to process the boards further. I personally don't. I like to sell them on eBay. They're going for about a dollar a piece right now. I personally have found a little more lucrative way of dealing with old cellphones. I found this company, Top Dollar Mobile, which they really are willing to pay top dollar prices for your phone you have. [She gives details about Top Dollar and how to sell to the phones.] So, for a phone like this, if it is working, they will give you forty bucks, and if it is not working, twenty. If I were to scrap this phone and get a board like this out of it, I would only get a dollar. A dollar, twenty to forty [dollars], you make the call. In this case, it might be worth it to save your scrapping tools for a bigger job. I'm Scrapper Girl – thanks for checking me out! (emphasis added)

This video is part of a YouTube video genre that details how to scrap or salvage electronics for the various precious and non-precious metals inside. While each video differs regarding the level of detail, all are highly performative and involve either an individual or a group speaking to the camera as they dismantle a device in a step-by-step process. As with all YouTube videos, comments abound; while ranging in their tone, viewers predominately appear to be thankful, seek clarification, or call out the scrapper for misinformation.

While engagements with e-waste are associated most spectacularly with sites in the Global South, such as Agbogbloshie in Ghana, the salvaging of electronics is also happening across North America at various scales (Lepawsky, 2018; Reno, 2015). Although the former have become part of the spectacle through which the detritus of our digital age are made visible in magazines, newspapers, and films (Hirsch, 2013), the latter are largely unseen save for the digital publics created through YouTube, eBay, and other digital forums. However, "urban mining" (as this practice is also known) is increasingly framed as an answer to the growing global e-waste problem, as a potential economic windfall, and as a cost-effective alternative to the mining of virgin deposits of precious metals such as gold and copper (Noyes, 2014; Gabrys, 2011; Zeng, Mathews, & Li, 2018).

The potential value of these devices is exemplified in a 2006 pamphlet by the U.S. Geological Survey (USGS) entitled "Recycled Cell Phones – A Treasure Trove of Valuable Metals." Assuming that a cellphone weighs 113 grams and that each of the 180 million mobile phone subscribers in the United States in 2004 used one phone, the USGS estimated that the collective assemblage of cellphones would be 20,000 metric tons. Of this total, the USGS calculated that 2,900 metric tons would be copper, 64 metric tons would be silver, 6 metric tons gold, "almost" 3 metric tons palladium, and 0.06 metric ton platinum (Sullivan, 2006). More recently, the global stream of e-waste – computers, smartphones, microwaves, televisions, refrigerators – has been projected to be 40 million metric tons (Breivik, Armitage, Wania, & Jones, 2014), with the further assessment that around US$21 billion in gold, silver, and other precious minerals are in our discarded electronic devices (Koestsier, 2013). In 2012, when the Scrapper Girl video was released, gold, the most recognized precious metal within these devices, was valued at US$1657.50 an ounce.[3] The real and potential value of the minerals within e-waste, combined with the decline in waged employment during the 2008 Great Recession in the United States and the increasing obsolesce of feature phones as people shifted to smartphones, created a particular niche for people with the time and inclination to take things apart in search of ways to make money.

Focusing on the virtually constituted community of scrappers through their digital publics (online forums, websites, and YouTube videos), I am interested in how expertise is performed and value materialized through scrapping and how scrapping entangles people in the lives of minerals. While this chapter helps illuminate the political economy of discarded electronics, these videos also give a unique perspective on the urban, and most often Global North, counterparts to the artisanal miners predominately from the Global South discussed elsewhere in this volume (especially in chapter two by Walsh and chapter five by Calvão). Despite their various differences, the scrappers discussed here are similarly bound by the need to make ends meet while pursing the "dirty" work of extracting otherwise hidden and precious minerals. Thinking through the performances of these women and men as they create digital publics of shared expertise, as well as the transforming materialities that scrappers are endeavouring to harness through their craft, I also touch on how the boundaries between self and object, and between truth and fiction, are blurred. Doing so, I demonstrate how, in re-enlivening our e-waste, scrappers are striving to remake themselves through their work and are thus bound up with minerals through their array of passions (see the introduction to this volume).

Through his reading of the French poet Baudelaire, Walter Benjamin's formulation of the ragpicker, an iconic figure of modernity (alongside the collector, flâneur, and detective), is a useful way to help understand these efforts at self-making. During the interwar years, the ragpicker was a perfect metaphor for the ways in which ideas, traditions, and things were constantly reconfigured and recycled as people sought to work out what modernity was and how to create new trajectories (Alexander & Reno, 2012: 8–10). These marginal individuals were key in transforming substances – pulling out and reassembling what value remained in societies' waste (Marx, 2007). In a similar manner, scrappers offer insight into the connections and valuations that are occurring on the margins of the economy in the wake of the Great Recession as they help transform and extract value from our collective electronic detritus. As is evident in the Scrapper Girl video, and as I will show in the other YouTube videos discussed later, scrappers collect what others neglect and then proclaim to possess not only the expertise to discern the hidden wealth within this material but also the ability to refine or sell what is valuable.

After a discussion of some theoretical considerations and broader economic contexts of scrapping, I turn to a set of scrappers who, through their YouTube performances, seek to enact different aspects of a digital public that coalesce around trajectories of hope, desire, and freedom. These performances range from the purely technical to highly theatrical and comic. Regardless of their style, each scrapper helps create a rhetoric of preciousness through which value is reinscribed into the various minerals within cellphones (gold, silver, and copper) and the possibilities that can be pursued of preciousness in other minerals (tantalum, lithium) are pointed out.

Scrapping or the Art of Revealing Assemblages

> Artefacts ... emerge – like forms of living beings – within the relational
> contexts of the mutual involvement of people and their environments.
> (Ingold, 2000: 88)

Following Ingold's assertion of the relational approach to objects, people, and the environment, I take the stance that, just as these relations emerge through making, they also emerge when objects are taken apart (Colloredo-Mansfeld, 2003; Hetherington, 2004; Graham & Thrift, 2007; Bell et al., 2018b). Scrapping is an active realization of what Jackson (2014) has termed broken world thinking – that is, an understanding that all objects will and do break and that breakdown helps to reveal the relations within technology otherwise obscured in the Global North (and elsewhere) by capitalism (Hornborg, 2001).[4] Though informed by my research in Papua New Guinea (Bell, 2008), my understanding of these relations is most centrally rooted in collaborative research on third party cellphone repair technicians in Washington, DC.[5] This research focused on how repair technicians reverse-engineered otherwise black-boxed technology to repair the phones for consumers (Bell et al., 2018b; Kuipers et al., 2018).

This collaborative research clarified how repair opens up perspectives on two aspects of cellphones, which can be productively applied to scrapping. The first is the realization of how commodities are "global assemblages" (Collier & Ong, 2005; Bell et al., 2018a). Drawing on Deleuze and Guattari (1987) and Haraway (1991; 1997), this formulation brings into view the materials, ideas, and labour that compose things, as well as the different scales involved and the transformations of value that take place in these formulations. This concept is part of recognizing what has been described as the "resource materialities" at play within all things (Richardson & Weszkalnys, 2014), which have a "distinctive history of formation but a finite life span" (Bennett, 2010: 24; see also Ferry, 2013; Gabrys, 2011; and other contributions in this volume). Minerals play a key – though often not recognized – role as part of the assemblages that inform our lives (see this volume's introduction).

A constant in these forays is the presence of minerals hidden within these electronic devices and the wealth that these minerals can become, which brings me to my second point: repair, as with scrapping, reveals – however momentarily – the supply chains that compose capitalism, by which these things come into being.[6] Drawing on Tsing's investigations of this aspect of capitalism as it relates to mushrooms (2009a; 2009b; 2015), repair functions as another node in which the supply chains loop back in on themselves in technicians' efforts to sustain and further a mobile phone's functional life for a consumer. In this liminal space, the value of the broken device is open to evaluation by repair technicians and consumers, and the technology is made livelier (Bell et al., 2018b). Scrapping critically diverges from repairing, as scrappers have no interest in making devices "function." Rather, the object being scrapped is a resource from which otherwise hidden and precious minerals can, with the right knowledge, be extracted for profit. However, both activities, through opening up the device in question, allow the individuals involved to pursue diverse ends (Tsing, 2015).

To this end, scrappers take the process of opening up devices a step further by dematerializing devices down into the minerals that make up their various components. While adding new phases to the biographies of the various materials that form an object (Kopytoff, 1986), scrappers, through the medium of YouTube, also assert the worth and skill of their labour in this transformation, which they frame as an enactment of the American dream of economic self-sufficiency or as an assertion of their technical knowledge (Tsing, 2015; Reno, 2016). In their ability to see value where others do not (not only in terms of content but also in scale), scrappers actualize in their practice Ingold's assertion that "every material is a becoming" (Ingold, 2012: 435; see also Bennett, 2010). Given the mutability of particular minerals found in cellphones – gold, silver, copper, and palladium – this becoming is particularly important as scrappers seek to extract and refine a device's components (Alexander & Reno, 2012; Walsh, 2004; Ferry, 2013).

The data used for this chapter are the multitude of videos hosted on YouTube that are produced by individuals (and companies), as well as conversations in online scrapper forums. These avenues are the main means through which "scrapper publics" are generated and sustained (Warner, 2002).[7] As cultural texts, these performative and instructional videos reveal attitudes and tensions in the work at hand and show how individuals are aiming to position themselves as experts and as part of a wider network of practitioners (Jones & Schieffelin, 2015).[8] This form of data lends itself readily to an analysis that examines, and highlights, how the linguistic and the material co-constitute one another in the creation and sustaining of value (Gershon & Manning, 2014; Shankar & Cavanaugh, 2012). The performative space of scrapping videos both exhibits and creates the expertise of the scrapper, while also working to legitimize the activity of scrapping.

At the heart of expertise are relations of ideology and power, involving the interactions between things, makers, and consumers of knowledge and the subsequent creation of "semi-stable hierarchies of value" through which expertise is enacted (Carr, 2010: 18). Expertise is thus performed and, as such, is embodied and linguistically enacted. But it is also temporary, so these performances must reoccur (Bourdieu, 1984). As a learned position, expertise is fundamentally collaborative, involving people with their objects of knowledge in a mutual constitution (Miller, 2005; see also the introduction in this volume). Expertise also demands intimacy, in that individuals become deeply familiar with the workings of particular technologies, practices, and/or institutions and develop material and linguistic repertories by which their knowledge is displayed and enacted. Expertise uses credibility to differentiate between those who know and those who don't, and is revelatory of the ideologies of those involved (Carr, 2010). Through these demonstrations of expertise, scrappers ideally drive traffic to their YouTube videos, establish a following, and supplement or earn an income.[9]

Some Notes on the Political Economy of Scrapping

While it is a truism that all recycling endeavours create secondary markets (Alexander & Reno, 2012), those of e-waste are tremendously varied. While the Basel Convention (which the United States has not yet ratified) has largely shaped how these materials are distributed globally (Lepawsky, 2012; 2015; 2018; Lepawsky & Billah, 2011; Lepawsky & Mather, 2011), within the United States varied state law and Environmental Protection Agency (EPA) regulations shape local distribution and availability. Here I focus on individuals whom I term artisanal scrappers (see chapter two), individuals who scrap as a hobby and as a living but who are not part of larger e-waste recycling and/or processing companies (Reno, 2016).

A 2010 article by *Urban Mining*, entitled "Striking Gold in Cell Phones," epitomizes the rhetoric around the real and imagined value of e-waste and gives some sense of how and why scrapping is profitable. As the article notes, cellphones

> hold some of the most valuable bang for the buck. Chock full of bits of gold, silver and copper, cell phones offer easy extraction and reuse of these valuable materials, and with technology rapidly advancing, there is no shortage in sight. Gold, which has rapidly increased in value during the past few years, is of particular interest to companies looking to cash in on e-waste extraction. (Urban Mining, 2010)

Citing gold's superior conductivity over copper, the article goes on to remark on the easily extractable gold in our mobile devices ("Per ton of recovered cell phone circuitry, it is possible to yield up to 150 grams [5.3 ounces] of gold.") and notes that this mineral is highly sought after by jewellers and electronic manufacturers. The article also cites the often commented upon "fact" that, with people unintentionally hoarding their electronics, "chances are good that a lot of precious metal is hiding in cabinets, drawers and closets around the world."[10]

While the details vary for each phone, the average feature phone consists of "40 percent metals, 40 percent plastic and 20 percent ceramics and trace materials" (EPA, 2004). Circuit boards or printed wiring boards contain copper, beryllium, gold, lead, nickel, and tantalum, while the board requires crude oil to make the plastic, sand, and limestone for fiberglass (EPA, 2004). Although the percentages and materials vary across phone models, scrappers consistently assert that it is more profitable to resell smartphones than to scrap them for their boards (see Breivik et al., 2014). Indeed, for most scrappers, the scale needed to extract enough precious minerals to make a substantial profit is out of reach; as a result, their scrapping is a way to earn an additional income while pursuing a hobby.

Scrappers are also typically quiet about the sources for their materials, though, as Moose Scrapper (discussed in the next section) comments to a viewer, "I get them by placing ads on craigslist and putting up signs in grocery stores. I offer free confidential recycling and destruction of personal info in electronics. I get a pretty good response."[11] These informally obtained cellphones involve a different set of relations than those bought and sold on eBay.

The Chaotic Passions of Scrappers

> Every passion borders on the chaotic. (Benjamin, 1968: 60)

While scrappers seem to cut across all ages, genders, and ethnicities, the community of YouTube scrappers in North America seems to be predominately white and working class. Each of the scrappers I discuss here – Scrapper Girl, Scrapperella, and Moose Scrapper – index aspects of this public. While all share a core of rhetorical and technical skills, they perform their expertise differently. In turn, they each reveal aspects of how minerals are integral to the creation of new prosperous selves and possess a deep passion for what they do.

To explain, let me return to Scrapper Girl, with whom I began this chapter. In her sixteen YouTube videos published over the space of a year (from 21 March 2012 to 7 April 2013), she covers a range of scrapping projects in addition to cellphones: how to scrap televisions and speakers, how to sort metal, and how to find and sell scrap. Indeed, for all the scrappers whose videos I viewed, cellphones are only one source among an array of electronics that they focus on. As with other scrappers, Scrapper Girl has an origin story, which she tells viewers while sitting casually on a set of stairs:

> I started scrapping after I went to visit my uncle in Pittsburgh and ... saw this really sweet rig he has – he fills about twice a week and when I was with him he went to cash it out and he made over 500 bucks! ... After I went with him to do that, I was totally hooked and I thought it was so neat. Eventually, I want to get my own rig and stuff ... but I am new at this [the video switches to shots of metal sheets as she continues to speak]. I am just starting out, figuring out different metals, grades, and that sort of stuff.

The video then jumps to a moustached man in a hard hat standing in a junkyard who smiles and says, "We like it when she comes in. She is pretty hot."[12]

While, Scrapper Girl's origin narrative echoes others, the video is distinct, not only through the final man's sexual framing of her but also because Scrapper Girl is an utter fabrication. As Jeff Lower, a cinematographer involved in the project, explains, Scrapper Girl was "a short series by Chinimblelore productions (I was hired as a camera operator) based on a girl who scraps metal for money. 16 episodes with over one million views."[13] Kevin Hicks, an owner of Hark Productions, which produced the series, provides more context:

> My partners and I have made some progress with a concept that we call "Scrapper Girl." It's a YouTube series that had met with some success and popularity on YouTube, has made us a little bit of money and has seen a few different TV people to reach out to us. Nothing has come of it yet, but we are dealing with a few interested parties. If you have an idea with some proprietary elements for instance (*Scrapper Girl is about an attractive woman who does "a man's job"*), then YouTube is good place to put those ideas. (Hicks, 2013; emphasis added)

Looking at the Scrapper Girl website, the full extent of the proprietary elements can be seen in the production company's store, wherein one can download a free wallpaper image of Scrapper Girl in a bikini standing in a scrapyard or purchase ringtones, T-shirts, mugs, posters, magnets, and keychains.[14] Here we have capitalism at its most perverse and in an obverse of what the ethos of scrapping aspires to be: generating wealth through dismantling others' discards. Although the webpage still stands, the items are no longer for sale.

Within the scrapper community, questions about Scrapper Girl's authenticity and knowledge quickly percolated up in commentary on her YouTube videos and in forums where the videos were posted.[15] These comments are instructive, as they reveal how expertise is policed and how passionate scrappers are about their activities. Take, for example, a thread labelled "Scrapper Girl" created by "sonicj" on the EEVblog Electronics Community Forum (1 June 2012), a forum self-described as "A Free & Open Forum for Electronics Enthusiasts & Professionals."[16] In response to sonicj's post, Bored@Work remarks, "Looks like staged to attract a certain young mail [*sic*] audience. Watch her hands. No scratches, no marks, no scars. She usually doesn't do such work with her hands." Others draw attention to Scrapper Girl's tools, pointing out their pristine state and the wrong way in which she holds them. These comments are suggestive of how this community sees their own expertise as constituted through their bodies, gestures, tools, and discursive knowledge.

It is Simon, however, a global moderator of the forum, who performs the fiercest boundary maintenance of the scrapping craft through his critique:

> What a load of garbadge [*sic*]. I could have taken that TV apart in the same time the video lasted (without the outake [*sic*] time) and with one tool: The hammer, the holy grail of dismantling for scrap. If you can't get it open and apart with a hammer it won't be worth your time ... *This sort of trash should be removed from the internet in the name of preserving resources for serious stuff like the eevblog.*[17] (30 June 2012; emphasis added)

This bravado works to reinscribe scrapping as a largely male-gendered activity, as do the frequent comments on Scrapper Girl's physique. At the same time that scrappers disavowed Scrapper Girl, some of them used her semi-celebrity to better position themselves within YouTube. I raise this point because a key underlying feature of these videos is to earn some margin of income through viewer traffic and ad revenue.

If Scrapper Girl is a simulacrum, my next example is a self-made scrapper who enjoys the iconoclasm of scrapping and the freedom it helps her achieve. Introducing a YouTube video entitled "Scrapper Girl – This Is How You Lift Metal! Chinimblelore – Junk Yard Scrapping SHOUT OUT," a scrapper named Scrapperella writes in the video's description:[18]

> Making $72.00 at the Metal Yard! I was inspired to give a shout out to Scrapper Girl due to the fact that [in] her last video she was struggling

with tipping a stove on to a hand truck. *Though it was amusing and she is easy on the eyes to watch. It is very apparent that she is more in to making videos about scrapping versus scrapping herself* ... Nothing wrong with working Smarter and Not Harder. *I just like the fact that I can be very independent when needed.*[19] (emphasis added)

The video itself opens with a brief monologue by Scrapperella, a white mid-thirties woman:

I was gonna put a video together, and I just happened to notice another YouTuber. I want to give a shout out to Scrapper Girl, and how do I say this ... She can't scrap by herself ... I can throw some metal around, so this is how it is done Scrapper Girl. But, seriously, guys and girls, the information on her YouTube account is informative, so go check her out and she's easy on the eye.

The rest of the video shows Scrapperella throwing metal off a truck, which is picked up by a magnet to get crushed. The video ends with Scrapperella commenting mockingly, "I am a girl, who can scrap [makes wink and ching sound]. If you've seen Scrapper Girl videos then you will know what that was all about." Establishing her own credibility through this theatrical performance of opposition (the video ends with her showing off the US$72 earned), Scrapperella nevertheless attempts to use the popularity of Scrapper Girl to drive more traffic to her own revenue-generating videos.

Scrapperella, as Elena Gonzalez calls herself, made her livelihood with her then husband Kenny "Big Hammer" through scrapping and other salvaging endeavours until their marriage ended in 2014. On YouTube since 2010, the pair have collectively made some 360 videos (as of 2017, all of the scrapping videos discussed here have been removed from Scrapperella's YouTube channel but can be accessed through the Internet Archive). Predating Scrapper Girl as she does, Scrapperella's main motivation behind her post appears to be frustration at the lack of authenticity. As she describes herself on her YouTube profile, "Scrapperella makes ends meet without the security of a paycheck by Ebaying, Youtubing, Scrapping, etc."[20] Further, in a comment to one of her videos, "How Lucrative Is Garbage Picking & Scrapping?" (20 October 2011), she notes:

What we do is Scrap, Picking (Which is Yard Sales, Trash, Junk Yard). Pretty much get what we can for cheap "Trash is Free" =) and turn around and sell it. *I was laid off in 2008, but was lucky enough to have unemployment for the first year to really start to get myself established.* Really took a lot of learning and in fact at that time I had two kids and a mortgage ... LOL, now it is three kids. Work from 7am to 4pm (M-F) and Kenny when he wants.[21] (emphasis added)

More revealing are comments she made in a YouTube video of an edited Google hangout with fans, entitled "Getting into Scrapping: Can You Make a Living? Does Everything Sell at Scrap Yard?" (19 February 2014). Responding to a question, Scrapperella remarks:

If you going to look at scrapping, you can look at it in many different ways. Some people just like gettin' dirty. Scrappin' [is] just something to do and the thought of getting paid for it, as a hobby. Recycling, saving the environment is enough for them, and they just enjoy scrapping. But they are not in it for the income. Then for me, someone like me [is] in it for an income, but if you are in it for the income, unless you have business flyers, full out scrapping vehicle and trailer ... the other business is to have other avenues to sell it whether you are selling it on flea market or eBay.[22]

Watching Scrapperella and her ex-husband's many videos and reading their commentaries, one gets a sense of the inexhaustibility of waste (Reno, 2016). What appears to be the challenge of all this waste is how to find ways to transform it efficiently into money. Reflecting further, Scrapperella comments:

But yeah it [scrapping] is almost an addiction. You know my nickname is Scrapperella and I love that nickname, and that's gonna carry on with me because that's where I started ... I'm always gonna like Scrapperella 'cause I don't wanna ever forget that I was out there making a living, going through things on the side of the road making it happen. I take pride in that. I love that part of my life. I'm thankful for it, it is quite an experience to do that and make it happen.[23]

Here scrapping emerges as a source of pride and freedom – as she and Big Hammer (her ex-husband's moniker) pronounce in the previously mentioned 2011 video, which is subtitled "We Are Living Our American Dream by Treasure Hunting in Trash!!!!" In keeping with their homemade videos, the piece has didactic intertitles with still images of treasures found. Following a shot of Big Hammer sitting on a stove in a dump truck with an "I heart scrap" sticker on the cab, overlaid with the question "What is the American Dream?," rolling intertitles proclaim: "Freedom! Freedom is the PROMISE of possibilities! Possibility of prosperity and success! Regardless of social class or circumstances of birth! All men are created equal!" "Life, Liberty and the pursuit of Happiness!" (compare Tsing, 2015: 82–106). The rest of the video asserts that treasures are to be found in trash and extols the virtues of scrapping in order to survive economic hardship.

The tone of their videos is playful and hinges largely on Big Hammer's prowess with his hammers in destroying things. Despite the wide-ranging scrapping activities, the pair only deal with cellphones in one video and gold in a handful of others. Their cellphone video entitled "Scrapping a Cellphone for Gold or Recycling It? Which Option PAYS Better?" appears in a series they call "Working Wednesday," in which they "demolish things for fun and for cash."[24]

The video begins with Big Hammer explaining that there is "one to four dollars' worth [of gold] in here," to which Scrapperella replies, "Really?" Big Hammer continues, "Yeah, we are going to find out." Then, declaring it "hammer time," he begins to smash a cellphone. Shifting away from the image of the hammer breaking the phone, the remainder of the video shows viewers what is in a cellphone through a panoramic of a deconstructed phone. The video suggests selling more expensive phones, such as BlackBerrys, through various vendors on eBay or other platforms. Despite destroying a phone in the video, Scrapperella and Big Hammer repeatedly stress that scrappers should follow the most cost-efficient and environmentally responsible method of disposal.

If Scrapperella is about materializing the American dream through scrapping and reflects the anxiety and financial instability caused by the Great Recession, Moose Scrapper, by contrast, is an artisan for whom scrapping is a passionate hobby. Beginning his YouTube channel on 17 October 2012, with over 52,706 subscribers to his channel as of 6 May 2019, he has produced 288 videos. In his introductory video, Moose Scrapper remarks:

> If you're watching this then that means you are trying to make some money … In this channel I have several videos. I am planning to post a whole lot more, showing you how to make money doing practically anything from scrapping electronics and stuff to selling things on eBay, *whatever it takes to make a few extra bucks and to make the bills a little extra at the end of the month.* So thanks for watching![25] (emphasis added)

Moose Scrapper has a more reserved personality and rarely discloses aspects of his personal life. He does, however, have Facebook and Google+ profiles through which he alternately circulates his videos. While Moose Scrapper's videos vary, he consistently demonstrates his desire to show viewers how to retrieve minerals from electronics.

Further, as of 2018, he is the only one of the three scrappers discussed still producing videos. The titles of his videos range from "Scrapping Motherboards, 24K Gold Motherboards and Types of Motherboards" to "How to Scrap a Hard Drive for Gold, Platinum, and Other Metals." The most technical of the scrappers discussed here, Moose Scrapper details the process of chemical gold extraction (acid peroxide method), the procedure to depopulate a board using a sand bath, and the way in which to smelt the resulting gold powder into a button.[26]

In his 14:26 minute video "Scrapping Cell Phones for Gold Recovery" (1 July 2014), Moose Scrapper begins by explaining that the video is offered in response to something he has had requests for. Looking over a timeline of phones, he notes that, after removing the batteries for recycling, a scrapper needs to remove the smart cards. Then, while opening and displaying a circuit board, he remarks:

> What I have learned on the gold refining forum is that these are actually fairly easy to get a yield or outcome measure from. It takes about a thousand of these to make one ounce of gold. I know you are thinking, one thousand – holy crap, that is a freakin' lot of cellphones ... but I've got 1, 2, 3, 4, 5, 6, 7 here and that is just from this week. So for all the cellphones you sell, take them out and set that aside and it will add up eventually. You will get there. Cellphones really don't have a hell of a lot of gold in them – it is really a myth that people think cellphones are loaded with gold. They are really not. Typically at most $4 in gold in a cellphone [taking them open]. So if you look at it [showing one board], there is not a whole heck of a lot of gold on there – there is some, but not much to get excited about. You can process the boards in acid peroxide, but before you do so, you want to get all of these things off in a sand bath and pick off these crystal oscillators here, those have gold in them. Monolithic ceramic capacitors, those will have palladium and silver. A lot of these little things will have gold on them. Make sure you put that aside for processing or selling elsewhere.[27]

While Moose Scrapper follows a similar convention to other scrappers when detailing how to open a cellphone, he gives a further level of detail about the cellphone's parts (and dispels popular ideas about cellphones being loaded with gold), which, alongside his refining videos, sets him apart. In this and other videos, he is careful to warn viewers about the toxicity of vapours and the caustic nature of materials used to extract gold. He also reminds viewers that they should conduct all their work in a well-ventilated space and with the proper safety equipment. But, if his expertise is on display in these videos, it also comes out in his detailed responses to watchers' comments. Here we begin to glimpse the wider community of scrappers and see how these individuals work collectively to understand what is inside their electronics and how they can extract these minerals to make money. Take the following examples from the comments Moose Scrapper has received to date on the video discussed earlier[28] and his responses to them:

DWIGHT GORDON: I've torn apart a few cell phones now. The stuff on the back of the white sheet that comes into contact with the gold plate (4:31) looks like silver. Is it? Some of the same stuff showed up in my wife's old non-working glucometer as well. Quite a bit of gold in those, by the way.

MOOSE SCRAPPER: I have tested some of those and have found them to not be silver.

EVILSEARCHENGINE: Hey Moose! Discovered you about 6 weeks ago and finally got up to date on your vids. Absolutely love 'em. My question regarding pins: The male pins are obvious, but how do you extract the hidden female pins from connectors in a timely manner? Possible vid?

MOOSE SCRAPPER: Thanks for watching! The timely manner part is what is difficult with the female ends. Honestly, I box them up and sell them by the pound. Once you get your name out there and people start to really reach out to you with scrap, it's hard to keep up and there is less time for the smaller time-consuming things. I think original feets has a video or two on some easy ways to remove those things with a simple petal pick. Thanks!

ANTHONYYWW713: Im just going to assume that smart phones or touch screens lack the gold since they were not featured. But awesome video btw.

MOOSE SCRAPPER: Smart phones and touch screen phone are worth much more money sold as is even if broken. WAY more than scrap price. I did not show a smart phone here because I sell them rather than scrap them to make the most money possible.

KEITH BILLS: I really enjoyed this video im new to scraping im trying to do it all stripping copper an learning about the cell phones an computers etc. i just figured if i went hardcore at it i could make a little money i lost my job of 7yrs an my wife left me an divorced me so this is stuff is perfect for getting things off.

MOOSE SCRAPPER: Thanks for watching, sorry to hear about your hard times. This is a great way to keep your mind occupied and make a little extra cash while doing it ... Thanks for watching!

Asking questions about tools, techniques, and style, Moose Scrapper's subscribers emerge as an engaged audience who, in some cases, reveal aspects of their personal life that led them to scrapping. These comments further reveal a community that has the time, interest, and apparent need for a way to earn money through scrapping. Moose Scrapper, for his part, answers everything calmly and with enthusiasm. Before concluding, I want to cite one last example from the comments, in which pacoblancosmith calls into question Moose Scrapper's expertise and his motivations for being on YouTube. His comments hint at how scrappers who use YouTube are also mining viewers as a way to earn more funds.[29]

> PACOBLANCOSMITH: As much as I enjoy watching your vids, and have been a subscriber for some time, this one was a bit lacking in fact. Plenty of opinion but not enough research and fact. I could show you some cell phone boards with a lot of gold plating on them. I do realize that once you have YouTube sending you a check every month based on your videos, it must be hard to stay true to your original drive and vision.
>
> MOOSE SCRAPPER: I am sorry you feel that way. I will reassess my videos more carefully before releasing them. I will however stand behind the statements about the amount of gold on cell phone boards. *There simply is not a whole lot coupled with the time it takes to deconstruct the phone, depopulate the board then process, as well as the availability of phones in volume, it is my opinion, as well as many others in the refining biz and on the goldrefiningforum that your better money will be in gold fingers off RAM chips, PCI cards, slot processors, etc.* While the boards may display a lot of gold surface area, that gold plating is extremely thin. I do sincerely thank you for the honest and constructive criticism though. That is infinitely more valuable than just calling me a sell out or something. Thanks again for taking the time to comment! (emphasis added)

Exhibiting his characteristic calm in this thorough response, Moose Scrapper's candor in this reply is notable. In the video itself, he debunks the myth that, in his words, "cellphones are loaded with gold" and tries to give pacoblancosmith a detailed answer about better sources of gold in electronics.

Conclusion

> The smaller the object, the more likely it seemed that it could contain in the most concentrated form everything else. (Arendt, 2007: 1)

By way of conclusion, I invoke Arendt's reflection on the work of Benjamin and his fascination with fragments. I do so not only to return to the figure of the ragpicker, with which I began this chapter, but also to invoke how cellphones and other electronics appear and are sources of precious minerals most often located elsewhere, as well as to caution that this chapter only provides a fragment of a much bigger social world. Our electronic detritus continues to elicit intense interest both for what it represents and for what value may lay within (Lepawsky, 2018). Scrappers, like Benjamin's ragpickers, and, indeed, artisanal miners and collectors discussed throughout this volume, are bound up with the potentialities of our collective detritus. Each is caught up in the allure of gold, alongside other minerals, found in our electronics. However, unlike the twentieth century ragpickers who, in Benjamin's words, "go about their solitary business" (2003: 48), the scrappers discussed here proclaim their actions through the digital domain of YouTube and other means. While broadcasting is part of their collective attempt to further extract revenue from their activities, it also speaks to the ways in which they are seeking to both educate their public and share their deep passions.

Although Scrapper Girl is the obvious anomaly in this discussion, being the fabrication of a production company looking to generate interest in its work, her videos reveal both perceptions of what scrapping entails and, perhaps more importantly, the ways in which the expertise of scrappers is defined and policed. Despite the fiction, the Scrapper Girl series generated interest in scrapping such that other scrappers responded to the work if only to help boost their own viewer numbers. By contrast, what emerges from the discussions of Scrapperella and Moose Scrapper is how they not only seek to demonstrate their skills but also strive to educate viewers on how to earn supplementary incomes through scrapping. The more performative of the two, Scrapperella, in her videos with her ex-husband, seeks to both entertain and educate. Her overt calls to the freedom of wresting treasure from trash openly point to an enduring ethos of self-reliance and ingenuity that scrappers celebrate. For his part, Moose Scrapper, through his technical videos, provides viewers with the techniques needed to achieve these ends while simultaneously pursuing his hobbies. Indeed, regardless of their intent, all of the scrappers are bound up with the passion that scrapping for precious and non-precious minerals elicits.

In this chapter, I have sought to illuminate a different set of connections that hover around cellphones and electronics: that of scrappers in North America. Focusing on the YouTube videos through which scrappers create a digital public and perform their expertise in the effort to educate each other and earn money, I have argued that the scrappers are also asserting their ability to be free and to transform *free* things into money. Indeed, while this chapter contributes to an understanding of the political economy of electronics (Gabrys, 2011; Lepawsky, 2018), it also points to an abiding tension in North America about the potential of finding hidden wealth (striking it rich) and living unfettered by financial worry (see Tsing, 2015; Reno, 2016). While the circumstances of the videos discussed here are intimately tied to the instability caused by the Great Recession of 2008, the ongoing production of videos by individuals such as Moose Scrapper points to something more. My sense is that scrappers persist in pursuing their passions because of the real and imagined possibilities that urban mining poses for both the desperate and the deeply interested. At one level, this ready source of materials enables women and men to pursue a hobby, while, on another, it allows them to explore real and imagined value contained within our discarded electronics, whose valuation only increases with time (Zeng, Mathews, & Li, 2018). While helping point to people's abiding interest in and entanglement with minerals, scrappers also reveal how they, and indeed we (though perhaps as not widely recognized), are deeply imbricated in the meshwork of our making and unmaking (Haraway, 1991; Ingold, 2012). Finally, scrappers help demonstrate how minerals, as essential aspects of the many things we use and consume, are generative, connecting those who engage with them in their various manifestations to an array of tangible and intangible knowledges about the world.

NOTES

1   Chinimblelore. (2012, 7 September). "Scrapper Girl scraps a cell phone for gold and talks about Top Dollar Mobile." *YouTube*. Retrieved from https://www.youtube.com/watch?v=mSQHSLTGYwA

2   For comparative purposes, see the number one video by IndeedItdoes, entitled "How to Scrap old Cell-Phones for *Gold Recovery," which was posted on 1 November 2011. The second video, "Scrapping Cell Phones for Gold Recovery," posted 1 July 2014 by Moose Scrapper, had 169,986 views and 218 comments on 15 September 2015. I will return to these videos later in my discussion, but note that I have focused on the videos I have due to the verbal exegesis of the scrapper. The IndeedItdoes video is a silent one in which white hands are seen opening and performing the scrapping while intertitles or textual overlays explain the process.

3   The year 2012 marked a historic high valuation for gold in the United States. In 2014, an ounce of gold was valued at US$1206.

4 While the knowledge of technology is varied, my contention is that, by and large, consumers in the Global North don't think about the material composition and supply chains through which their things are made. While there are obvious counter-examples found in the do-it-yourself (DIY) and ethical consumption movements, these issues are aspects of an object's "humility" (Miller, 1987; Bell et al., 2018a).

5 With support from a Smithsonian Consortia Grant for a project entitled *Fixing Connections: The Art & Science of Repair*, Joel Kuipers (George Washington University) and I led a research team composed of Jacqueline Hazen (PhD student at New York University), Briel Kobak (PhD student at the University of Chicago), and Amanda Kemble (PhD student at the University of Michigan) on cellphone repair over the course of eighteen months (2012–2014).

6 Different nodes of cellphone supply chains have come into view as anthropologists have examined the labour and social worlds of extraction (Smith & Mantz, 2006; Mantz, 2008; 2018; Smith, 2011; 2015), manufacture (Ngai, 2005; Smith & Mantz, 2006; Lüthje et al., 2013), infrastructures related to the use of mobile telephony (Parks & Schwoch, 2012; Parks & Starosielski, 2015), and the impacts of their disposal (Lepawsky & Billah, 2011; Tong & Wang, 2012; Lepawsky, 2015; 2018).

7 YouTube videos are one platform within a wider array of media through which individuals communicate and sell their materials. While they vary, these platforms always include Facebook, Twitter, and eBay.

8 The scrappers I discuss here are similar to the technicians examined by Orr (1996), with the difference being that, for the scrappers, their public is digitally constituted and the stories they tell about cellphones and techniques have a wider circulation.

9 For the mechanics of YouTube, see Burgess & Green, 2009.

10 A self-purported clearing house for information on the "process of reclaiming compounds and elements from products, buildings, and waste," *Urban Mining* set out ten steps outlining how the process works, each replete with the rhetoric of money waiting to be made (retrieved from https://web.archive.org/web/20170708030117/http://urbanmining. org/how-urban-mining-works) (1) "You might not know it but there are veins of precious minerals that are richer than any goldmine, running through our cities"; (2) "You might imagine that your new phone will last forever ..."; (3) "... or [after] a nasty accident, you're ready to be rid of it"; (4) "You don't want to throw it away ..."; (5) "So you hold onto it"; (6) "However, it's easy enough to recycle – and it's worthwhile to do so"; (7) "The devices are then disassembled ..."; (8) "And broken into their component materials"; (9) "From which the materials can be extracted

and reused ..."; (10) "... and could even be turned into the luxuries of tomorrow."

11 Moose Scrapper. (2014, 1 July). "Scrapping cell phones for gold!" *YouTube*. Retrieved from https://www.youtube.com/watch?v=miTNS7F2Xc8

12 This video is entitled "Scrapper Girl talks about how she got her start in scrapping." It was posted on 6 August 2012 and, by 1 May 2015, had 20,321 views. Retrieved from https://www.youtube.com/watch?v=F9TF-XbOn6E

13 Jeff Lower. (n.d.). "Web series," *Jeff Lower Video*. Retrieved from http://www.jefflowervideo.com/#/web-series/

14 Chinimblelore. (n.d.). "Scrapper Girl." Retrieved from http://www.chinimblelore.com/scrapper/store/index.html

15 The only positive review I located was by "jackshmuc," or Jack the Scrapper, who, in the Scrap Metal Forum ("The Internet's #1 Metal Recycling Forum"), created a thread entitled "Hot scrapper girl actually knows her stuff" in which he comments, "IF YOU HAVEN'T SEEN HER YOU ARE MISSING OUT ON SOME GOOD YOUTUBE and don't forget to check out some of her other videos and comment letting her know who sent you" (posted on 16 May 2012 on the forum, http://www.scrapmetalforum.com/scrap-metal-videos/9761-hot-scrapper-girl-actualy-knowsher-stuff.html). A scrapper and Ontario YouTube personality, Jack the Scrapper seems to be motivated more through self-interest in extolling her work – that is, he is trying to leverage her views into his own audience, branding, and ad-driven revenue. Acknowledging his support, Scrapper Girl produced a 34 second video entitled "Scrapper Girl Says Thanks to Jack the Scrapper – Shout Out!" and posted it to YouTube on 16 May 2012. Retrieved from https://www.youtube.com/watch?v=1y2aYVOm_NI&index=14&list=PLLa0vMZGgUR6ISgUeawra5GkwAlZEAU8-

16 The forum describes itself as "A Free & Open Forum For Electronics Enthusiasts & Professionals." The Scrapper Girl thread can be found at http://www.eevblog.com/forum/chat/scrapper-girl/

17 See the Scrapper Girl thread at http://www.eevblog.com/forum/chat/scrapper-girl/. This dismissive critique on the forum is not only reserved for Scrapper Girl but is also leveled at other YouTube personalities.

18 Since beginning this research, Scrapperella's videos are increasingly harder to access.

19 The video was posted on 16 October 2012 at https://www.youtube.com/watch?v=E6ELEzsT4OU, and I accessed it on 1 May 2015. However, the video is no longer available on YouTube, but can be found archived here: https://web.archive.org/web/20130203172133/http://www.youtube.com/watch?v=E6ELEzsT4OU&gl=US&hl=en

20 For Scrapperella's profile, see https://www.youtube.com/user
    /KennyChumsky/about
21 Scrapperella. (2011, 20 October). "How lucrative is garbage picking &
    scrapping?" *YouTube*. Retrieved on 1 May 2015 from https://www.youtube
    .com/watch?v=6m7BhkibHSg. This video is no longer available.
22 Scrapperella. (2014, 19 February). "Getting into scrapping: Can you make
    a living? Does everything sell at scrap yard?" *YouTube*. Retrieved on 1 May
    2015 from https://www.youtube.com/watch?v=oCBhNsRMgGE
23 In reference to this video, Frank Boston commented, "You should take
    pride. I love your positive attitude and can tell how proud you are of
    yourself. You made shit happen! It's so easy to live off the government and
    handouts. Scrapperella is a worker. One that is able to do what she loves
    and pay the bills doing it. This is the American Dream. You're on top of
    the world and you are Scrapperella ... forever! Thanks for all your videos."
24 This video was posted 18 July 2012, and on 1 May 2015 had received
    19,012 views. Retrieved on 1 May 2015 from YouTube. This video can be
    found archived here: https://web.archive.org/web/20120811232947
    /https://www.youtube.com/watch?v=0QGQw4bNQcw
25 Moose Scrapper. (2013, April 19). No title. Retrieved on 1 May 2015 from
    https://youtu.be/pdDDDllG66k
26 Discussing motherboards, Moose Scrapper details the value of different
    boards and the materials found in each component. Regarding tantalum
    capacitors, he notes, "Things have blown up on my Facebook page with
    people wanting to know about these things ... I have no idea how to re-
    cover that; I just remove the components in bulk and sell them. It is one of
    those materials that is expensive and has many interesting properties ... As
    far as processing anything else, you can almost process almost anything
    with those three methods: AP, hot HCL or incineration."
27 Moose Scrapper. (2014, 1 July). "Scrapping cellphones for gold recovery."
    YouTube. Retrieved on 1 May 2015 at https://www.youtube.com
    /watch?v=miTNS7F2Xc8&t=18s
28 The comments cited can be read here: Moose Scrapper. (2014, 1 July).
    "Scrapping cellphones for gold!" *UAClips*. Retrieved on 1 May 2015 from
    https://uaclips.com/video/miTNS7F2Xc8/scrapping-cell-phones-for
    -gold-recovery.html
29 Other scrappers often praise each other, and then ask to be checked out.
    As Vaughan Scrapper comments on this video, "Hey Moose great videos!
    I just made a youtube account and i am a scrapper as well :) would you
    mind checking out my channel? Thanks and keep up the vids!" To which
    Moose Scrapper comments, "Great job buddy! I'm gonna send some view-
    ers your way!"

REFERENCES

Alexander, C., & Reno, J. (Eds.). (2012). *Economies of recycling: The global transformation of materials, values and social relations*. London: Zed Publications.

Arendt, H. (2007). Introduction. In W. Benjamin, *Illuminations*. New York: Random House.

Bell, J.A. (2008). Promiscuous things: Perspectives on cultural property through photographs in the Purari Delta of Papua New Guinea. *International Journal of Cultural Property*, *15*(2), 123–39. https://doi.org/10.1017/S0940739108080107

Bell, J.A., Kobak, B., Kuipers, J., & Kemble, A. (2018a). Introduction. Unseen connections: The materiality of cell phones. *Anthropological Quarterly*, *91*(2), 465–84. https://doi.org/10.1353/anq.2018.0023

Bell, J.A., Kuipers, J., Hazen, J., Kemble, A., & Kobak, B. (2018b). The materiality of cell phones repair: Re-making commodities in Washington, D.C. *Anthropological Quarterly*, *91*(2), 603–33. https://doi.org/10.1353/anq.2018.0028

Benjamin, W. (1968). *Illuminations*. New York: Random House.

Benjamin, W. (2003). *Selected Writings: Vol. 4. 1938–1940*. M.W. Jennings & H. Eiland (Eds.). Cambridge, MA: Harvard University Press.

Bennett, J. (2010). *Vibrant matter: A political ecology of things*. Durham, NC: Duke University Press.

Bourdieu, P. (1984). *Distinction: A social critique of the judgment of taste*. London: Routledge & Kegan Paul.

Breivik, K., Armitage, J.M., Wania, F., & Jones, K.C. (2014). Tracking the global generation and exports of e-waste: Do existing estimates add up? *Environmental Science & Technology*, *48*(15), 8735–43. https://doi.org/10.1021/es5021313

Burgess, J., & Green, J. (2009). *YouTube: Online video and participatory culture*. Cambridge: Polity Press.

Carr, E.S. (2010). Enactments of expertise. *Annual Review of Anthropology*, *39*(1), 17–32. https://doi.org/10.1146/annurev.anthro.012809.104948

Collier, J.S., & Ong, A. (2005). Global assemblages, anthropological problems. In A. Ong & J.S. Collier (Eds.), *Global assemblages: Technology, politics, and ethics as anthropological problems* (pp. 3–21). Malden, MA: Blackwell Publishing.

Colloredo-Mansfeld, R. (2003). Introduction: Matter unbound. *Journal of Material Culture*, *8*(3), 245–54. https://doi.org/10.1177/13591835030083001

Deleuze, G., & Guattari, F. (1987). *A thousand plateaus: Capitalism and schizophrenia*. B. Massumi (Trans.). Minneapolis: University of Minnesota Press.

Environmental Protection Agency (EPA). (2004). *The life cycle of a cell phone*. Retrieved from https://nepis.epa.gov/Exe/ZyPDF.cgi/P100CAA7.PDF?Dockey=P100CAA7.PDF

Ferry, E.E. (2013). *Minerals, collecting, and value across the U.S.-Mexican border*. Bloomington: University of Indiana Press.

Gabrys, J. (2011). *Digital rubbish: A natural history of electronics*. Ann Arbor: University of Michigan Press.

Gershon, I., & Manning, P. (2014). Language and media. In N.J. Enfield, P. Kockelman, & J. Sidnell (Eds.), *The Cambridge handbook of linguistic anthropology* (pp. 559–76). Cambridge: Cambridge University Press.

Graham, S. & Thrift, N. (2007). Out of order: Understanding repair and maintenance. *Theory, Culture & Society*, 24(3), 1–25. https://doi.org/10.1177/0263276407075954

Haraway, D. (1991). *Simians, cyborgs and women: The reinvention of nature.* New York: Routledge.

Haraway, D. (1997). *Modest_witness@second_millennium.femaleman©meets_oncomouse™: Feminism and technoscience*, New York: Routledge.

Hetherington, K. (2004). Secondhandedness: Consumption, disposal, and absent presence. *Environment and Planning D: Society and Space*, 22(1), 157–73. https://doi.org/10.1068/d315t

Hicks, K. (2013, 30 October). The challenges of filmmaking and video production. *DegreeStory*. Retrieved on 1 May 2015 from www.degreestory.com/browse/bachelor/arts-religion/liberal-arts/the-challenges-of-filmmaking-and-video-production. This website is no longer available.

Hirsch, A. (2013, 14 December). "This is not a good place to live": Inside Ghana's dump for electronic waste. *The Guardian*. Retrieved from https://www.theguardian.com/world/2013/dec/14/ghana-dump-electronic-waste-not-good-place-live

Hornborg, A. (2001). Symbolic technologies: Machines and the Marxian notion of fetishism. *Anthropological Theory*, 1(4), 473–96. https://doi.org/10.1177/14634990122228854

Ingold, T. (2000). *The perception of the environment: Essays on livelihood, dwelling and skill*. London: Routledge.

Ingold, T. (2012). Toward an ecology of materials. *Annual Review of Anthropology*, 41(1), 427–42. https://doi.org/10.1146/annurev-anthro-081309-145920

Jackson, S.J. (2014). Rethinking repair. In T. Gillespie, P. Boczkowski, & K. Foot (Eds.), *Media technologies: Essays on communication, materiality and society* (pp. 221–40). Cambridge, MA: MIT Press.

Jones, G.M., & Schieffelin, B.B. (2015). The ethnography of inscriptive speech. In R. Sanjek & S. Tratner (Eds.), *eFieldnotes: Makings of anthropology in a digital world* (pp. 210–30). Philadelphia: University of Pennsylvania Press.

Koestsier, J. (2013, 5 March). Urban mining: Recovering $21B a year in gold and silver from discarded devices (infographic). *Venturebeat*. Retrieved from https://venturebeat.com/2013/03/05/urban-mining-recovering-21b-a-year-in-gold-and-silver-from-discarded-devices-infographic/

Kopytoff, I. (1986). The cultural biography of things: Commoditization as process. In A. Appadurai (Ed.), *The social life of things: Commodities in cultural perspective* (pp. 64–91). Cambridge: Cambridge University Press.

Kuipers, J., & Bell, J.A, with Kobak, B., Kemble, A., & Hazen, J. (2018). Intimate materialities in cell phone repair: Performance, anxiety and trust in DC repair shops. In J.A. Bell & J. Kuipers (Eds.), *Linguistic and material intimacies of cell phones* (pp. 237–65). London: Routledge Press.

Lepawsky, J. (2012). Legal geographies of e-waste legislation in Canada and the US: Jurisdiction, responsibility and the taboo of production. *Geoforum*, *43*(6), 1194–1206. https://doi.org/10.1016/j.geoforum.2012.03.006

Lepawsky, J. (2015). The changing geography of global trade in electronic discards: Time to rethink the e-waste problem. *The Geographical Journal*, *181*(2), 147–59. https://doi.org/10.1111/geoj.12077

Lepawsky, J. (2018). *Reassembling rubbish: Worlding electronic waste*. Cambridge, MA: MIT Press.

Lepawsky, J., & Billah, M. (2011). Making chains that (un)make things: Waste-value relations and the Bangladeshi rubbish electronics industry. *Geografiska Annaler: Series B, Human Geography*, *93*(2), 121–39. https://doi.org/10.1111/j.1468-0467.2011.00365.x

Lepawsky, J., & Mather, C. (2011). From beginnings and endings to boundaries and edges: Rethinking circulation and exchange through electronic waste. *Area*, *43*(3), 242–9. https://doi.org/10.1111/j.1475-4762.2011.01018.x

Lüthje, B., Hürtgen, S., Pawlicki, P., & Sproll, M. (2013). *From Silicon Valley to Shenzhen: Global production and work in the IT industry*. Lanham, MD: Rowman & Littlefield.

Mantz, J.W. (2008). Improvisational economies: Coltan production in the Eastern Congo. *Social Anthropology*, *16*(1), 34–50. https://doi.org/10.1111/j.1469-8676.2008.00035.x

Mantz, J.W. (2018). From digital divides to creative destruction: Epistemological encounters in the regulation of the "blood mineral" trade in the Congo. *Anthropological Quarterly*, *91*(2), 525–49. https://doi.org/10.1353/anq.2018.0025

Marx, U. (Ed.). (2007). *Walter Benjamin's archive: Images, texts, signs*. New York: Verso.

Miller, D. (1987). *Material culture and mass consumption*. Oxford: Blackwell.

Miller, D. (2005). Materiality: An introduction. In D. Miller (Ed.), *Materiality* (pp. 1–50). Durham, NC: Duke University Press.

Ngai, Pun. (2005). *Made in China: Women factory workers in a global workplace*. Durham, NC: Duke University Press.

Noyes, K. (2014, June 26). Can "urban mining" solve the world's e-waste problem? *Fortune*. Retrieved from http://fortune.com/2014/06/26/blueoak-urban-mining-ewaste/

Orr, J. (1996). *Talking about machines.* Ithaca, NY: Cornell University Press.

Parks, L., & Schwoch, J. (Eds.). (2012). *Down to Earth: Satellite technologies, industries, and cultures.* New Brunswick, NJ: Rutgers University Press.

Parks, L., & Starosielski, N. (Eds.). (2015). *Signal traffic: Critical studies of media infrastructures.* Urbana: University of Illinois Press.

Reno, J. (2015). Waste and waste management. *Annual Review of Anthropology, 44*(1), 557–72. https://doi.org/10.1146/annurev-anthro-102214-014146

Reno, J. (2016). *Waste away: Working and living with a North American landfill.* Oakland: University of California Press.

Richardson, T., & Weszkalnys, G. (2014). Introduction: Resource materialities. *Anthropological Quarterly, 87*(1), 5–30. https://doi.org/10.1353/anq.2014.0007

Shankar, S., & Cavanaugh, J.R. (2012). Language and materiality in global capitalism. *Annual Review of Anthropology, 41*(1), 355–69. https://doi.org/10.1146/annurev-anthro-092611-145811

Smith, J.H. (2011). Tantalus in the digital age: Coltan ore, temporal dispossession, and "movement" in the Eastern DR Congo. *American Ethnologist, 38*(1), 17–35. https://doi.org/10.1111/j.1548-1425.2010.01289.x

Smith, J.H. (2015). "May it never end": Price wars, networks, and temporality in the "3 Ts" mining trade of the Eastern DR Congo. *HAU: Journal of Ethnographic Theory, 5*(1), 1–34. https://doi.org/10.14318/hau5.1.002

Smith, J.H., & Mantz, J.W. (2006). Do cellular phones dream of civil war? The mystification of production and the consequences of technology fetishism in the Eastern Congo. In M. Kirsch (Ed.), *Inclusion and exclusion in the global arena* (pp. 71–93). New York: Routledge.

Sullivan, D.E. (2006). *Recycled cell phones – A treasure trove of valuable metals.* U.S. Geological Survey Fact Sheet 2006-3097. Retrieved from https://pubs.usgs.gov/fs/2006/3097/fs2006-3097.pdf

Tong, X., & Wang, J. (2012). The shadow of the global network: e-waste flows to China. In C. Alexander & J. Reno (Eds.), *Economies of recycling: Global transformations of material values and social relations* (pp. 98–118). London: Zed Books.

Tsing, A. (2009a). Beyond economic and ecological standardisation. *The Australian Journal of Anthropology, 20*(3), 347–68. https://doi.org/10.1111/j.1757-6547.2009.00041.x

Tsing, A. (2009b). Supply chains and the human condition. *Rethinking Marxism, 21*(2), 148–76. https://doi.org/10.1080/08935690902743088

Tsing, A. (2015). *The mushroom at the end of the world: On the possibility of life in capitalist ruins.* Princeton, NJ: Princeton University Press.

Urban Mining. (2010). Striking gold in cell phones. *Urban mining.* Retrieved from archive at https://web.archive.org/web/20100712195444/http://urbanmining.org/2010/06/03/urban-mining-striking-gold-in-cell-phones/

Walsh, A. (2004). In the wake of things: Speculating in and about sapphires in Northern Madagascar. *American Anthropologist*, *106*(2), 225–37. https://doi.org/10.1525/aa.2004.106.2.225

Warner, M. (2002). Publics and counterpublics. *Public Culture*, *14*(1), 49–90. https://doi.org/10.1215/08992363-14-1-49

Zeng, X., Mathews, J., Li, J. (2018). Urban mining of e-waste is becoming more cost-effective than virgin mining. *Environmental Science & Technology*, *52*(8), 4835–41. https://doi.org/10.1021/acs.est.7b04909

2 What Is "Artisanal" about "Artisanal Mining"? Reflections from Madagascar's Sapphire Trade

ANDREW WALSH

Although the activity known as "artisanal and small-scale mining" (ASM) has been around for millennia, the conventional use of this phrase to denote "low-tech, labour-intensive mineral extraction and processing" (Hilson & McQuilken, 2014: 1) is fairly new. More precisely, the "artisanal" nature of this distinctive sort of work, previously known simply as "small-scale mining," has only recently been specified by observers. While there are critical questions to be asked about how and why this label has come to stick, in this chapter I am more concerned with its appropriateness as a means of denoting a particular kind of work. Drawing from recent research and reflections on artisanal production generally and artisanal mining specifically, as well as from my own ethnographic research with artisanal sapphire miners in Madagascar, I focus especially on what might be gained from attending more explicitly to the "artisanal" nature of "artisanal mining" in our analyses of this activity.

In a recent review of research and reports on ASM in sub-Saharan Africa, Hilson and McQuilken discuss the development and systematic marginalization of this important sector of the global economy. How is it, they ask, that an activity that has come to occupy and affect so many millions of people around the developing world can "continue to be overlooked in most international, regional and local economic policies and programs?" (2014: 1). When ASM *is* referenced in the foundational conferences and reports behind influential understandings of this sector, attention tends to focus on the inefficiencies and problems associated with the particular kinds of work it involves. One report, for example, defines artisanal mining as "manual and very labour-intensive" mining carried out "by individuals, groups, families, or cooperatives with minimal or no mechanization, often informally and/or illegally" (Dorner, Franken, Liedtke, & Sievers, 2012: 1). Another describes

ASM as involving low levels of "occupational safety," "health care," and "productivity," and as a sector that provides minimal income to poorly qualified personnel engaged in inefficient work (Hentschel, Hruschka, & Priester, 2003: 6). The social and ecological contexts in which ASM takes place are also commonly portrayed as deficient, characterized both by what they lack – "social security" and "working and investment capital," for example (Hentschel et al., 2003: 6) – and by the harms they enable, including "prostitution, disease and narcotics consumption," "child labour," and environmental degradation (Hilson & McQuilken, 2014: 2).

Given the image suggested by such characterizations of ASM, readers might reasonably assume that there are more pressing questions to ask than what is "artisanal" about it. Indeed, some observers have noted how the convention of referring to such work as "artisanal" is problematic, given what Lahiri-Dutt terms the (clearly unwarranted) "ennobling connotation" (Lahiri-Dutt, 2014: 13) of this label. From such a perspective, the "artisanal" in "artisanal mining" masks or obfuscates the true nature of the work to which it refers. As journalist Geoffrey York notes, one way of reading or hearing the phrase "artisanal mining" in light of the often brutal circumstances faced by those who do it is as "a bureaucratic euphemism for the job of scavenging, digging and clawing a living from the harsh earth with bare hands and crude tools" (York, 2012).

Although I am as wary as any critical observer of the distracting power of well-chosen words, I am also conscious of the value of ideas that can help to refine how we think about complex phenomena. What follows is thus premised on the conviction that anyone concerned with ASM might come to understand this sector better by considering how the work to which the phrase "artisanal mining" has come to apply is, in fact, "artisanal" in more than just name.

"Artisans" in Anthropology and Beyond

It is understandably disconcerting to some that small-scale miners around the world have come to be known to analysts and researchers as "artisanal" at around the same time as have the bakers, chocolatiers, and cheesemakers favoured by a growing niche of worldly, socially conscious, and relatively affluent consumers. Clearly, the men and women I have encountered in northern Madagascar's gemstone and gold-mining communities over the past fifteen years have little in common with the specialty vendors I am likely to meet on a Saturday morning visit to my local farmer's market or a craft fair. But just *how* are they different? In this section I consider how twenty-first century "artisanal production" – defined by Colloredo-Mansfeld as "small-scale, minimally capitalized commodity production and sales" (2002: 115) – commonly engages differently positioned people in similar constellations of relationships that entangle them with the materials of their work and with other people simultaneously and in distinctive ways. I draw especially from recent anthropological considerations of artisans as, on the one hand, timeless and exemplary human actors whose engagements with their material environments offer food for thought on the limits of human intentionality and, on the other, as situated people engaged in distinctive relationships with others and with local and global markets for what they produce.

In the work of anthropologists and others intent on drawing attention to the role that non-human agents play in shaping human lives, artisans have commonly been portrayed as exemplars of a particular way of knowing and being in the world. Ingold (2012), for example, refers to the work of Simondon, Deleuze, Guattari, and others in describing how brick-makers, metallurgists, and other "artisans" engage in work that leads them to distinctive understandings of the influential "vibrancy" (Bennett, 2010) of non-human matter. Artisans, Ingold argues, know better than to imagine that they can simply impose their designs and intentions on the material world. Rather, they learn to "follow the flow" (2012: 433) of the materials of their work, guided always by "intuition" gained through hands-on experience. "In the act of production," Ingold writes, "the artisan couples his own movements and gestures – indeed, his very life – with the becoming of his materials, joining with and following the forces and flows that bring his work to fruition" (2012: 435).

In one sense, the exemplary artisan found in accounts like Ingold's is fairly generic; representing the maker in all of us, the artisan engages in a process that involves "not the imposition of preconceived form on raw material substance, but the drawing out or bringing forth of potentials immanent in a world of becoming" (Ingold, 2013: 31). What makes artisans special, however, is that their work predisposes them to a profound awareness of the limits of their own mastery of the material world; "[i]t is the artisan's desire to see what ... material can *do*," Ingold argues, "by contrast to the scientist's desire to know what it *is*" (2013: 31). Taken a little further, Ingold's perspective on the work of artisans invites us to reflect on the extent to which the variability of artisanal (or any) production (and, by extension, of the specific social, political, and economic arrangements it involves in particular contexts) stems from factors beyond human control. "Making," Ingold writes, "is a process of correspondence" (2013: 13); exemplified most obviously in the work of artisans, it is a process involving back-and-forth relationships that shape human and non-human correspondents alike.

While Ingold's reflections on artisanal production offer valuable insights on how non-human materials influence the lives of the humans who correspond with them, the timeless, decontextualized, ideal artisanal types represented in this work can seem far removed from the actual cobblers, weavers, cheesemakers, and other artisans of our times. Recent ethnographic studies of artisans in the contemporary world offer a different perspective, presenting artisans as people engaged not just with the materials of their work but also with the tools of their trades, with competitors, with the markets for which they produce, with state and international authorities, and, most generally, with the trends and forces of local and global political economies. In fact, while some observers, authorities, and consumers might imagine artisans as guardians of "tradition" and their work as a "survival" of pre-modern times, recent ethnographic accounts portray artisans as pioneers of myriad and profoundly uncertain late twentieth and early twenty-first century contexts, improvising new niches of commodity production and trade that are enabled by, though often at odds with, the forces of global neoliberal capitalism. Such artisanal producers are more numerous than many imagine. While it is true that artisanal commodity production for local markets around the world has declined over recent decades as "[m]ass produced, standardized and cheap factory items have replaced many of the various goods once produced by ... artisans" (Scrase, 2003: 449), artisanal production for *global* markets is anything but scarce. That in mind, it seems short-sighted to characterize artisans and their work as "pleasant ornaments to the globalized economy" (Giordano, 2002: 127), or to dismiss ethnographic research on artisans as descriptions of "small" or "cozy worlds" (128) of disconnected others. Studying living artisans may never provide us with a window onto the past, but it does offer a distinctive vantage on the workings of the contemporary global political economy.

Several recent ethnographic studies of artisanal production illustrate how artisans make do amidst circumstances in which their work is simultaneously highly valued and marginalized. Herzfeld's research with Greek artisans (2004), for example, reveals workers whose seeming incompatibility with the "modern" world is precisely what leads powerful others to put them on a pedestal that both exposes and limits them. The cobblers, carpenters, and other artisans with whom Herzfeld did research in Crete are "charged" by powerful political actors "with the business of reproducing tradition," oftentimes in a way they come to embody by "utilizing premechanical methods" (2004: 199). But they do so in a way that Herzfeld sees as keeping them in place – chained to the pedestal on which they are raised – "on the margins of a modernity already co-opted by the emergent and rapidly growing middle class, itself hostage to concentric expectations on the larger stage of international relations" (199). Not that this marginal status marks artisans off as anomalous and irrelevant. For Herzfeld, the ethnographic study of artisans offers a distinctive, bottom-up, perspective on "the operation of the global hierarchy of value" (27) that affects us all.

Colloredo-Mansfeld (2002) also portrays artisans as exposed actors on the cutting edge of forces that shape the global political economy. Remarking on how artisan trades have surged over recent decades, especially "in places that have embraced promarket reforms and global integration" (115), he focuses specifically on the pernicious ways in which such trades shape artisans' work and social lives. Regardless of what foreign consumers might imagine, the woven belts and paintings produced by the Ecuadorean artisans on whom he focuses are as much products of the contemporary global political economy as they are of tradition, and the economies that develop around the production and trade of these commodities are, not surprisingly, rife with competition, excess, risk, and other hallmarks of capitalism (Antrosio & Colloredo-Mansfeld, 2014). Artisanal production in Ecuador, and elsewhere, is not just a subset of global capitalist production, however. As Colloredo-Mansfeld notes, citing Garcia Canclini (1995), many late twentieth and twenty-first century artisans are notable for how they have avoided "rationalizing management, accumulating capital, converting to 'free' labor, mechanizing" (Colloredo-Mansfeld, 2002: 124), and other trappings of "modern" production, and in so doing have forged productive niches characterized by opportunities of a sort that large-scale, rationalized, capital-intensive, high-tech, industrial projects could never provide.

In discussing what is "artisanal" about artisanal cheesemaking in the United States, Paxson offers another example of how artisanal production is "inescapably defined against the industrial" (2012: 128). What makes artisanal cheese "artisanal" in the eyes of those who value it as such, she argues, is that it is "made more by hand than by machine, in small batches compared to industrial scales of production, using recipes and techniques developed through the practical knowledge of previous artisans rather than via the technical knowledge of dairy scientists and industrial engineers" (128). In addition to illustrating how American artisanal cheesemakers are, like the artisans described by Herzfeld and Colloredo-Mansfeld, caught up in webs of relations that entangle them with fellow artisans, with others further along the commodity chain, with consumers, and with the authorities who regulate their production, Paxson also shows how these producers are artisans of a sort that Ingold might recognize. The artisan cheesemakers she interviewed "consistently describe what they do ... as a balance of art and science" (129), that is, as involving "creative expression" and "an intuitively interpretive grasp of [their] materials" as well as "empirical observation and measurement, disciplined attention to record-keeping, and steps taken to ensure product-safety" (129). Here again, even as a contrast between "artisanal" and "modern" logics and methods of production are clearly drawn, there is no doubt that, as Paxson puts it, "the ethos of the contemporary artisan is not a throwback to the past; it is a modern pastiche" (142).

Like Paxson, I see great promise in approaching artisanal production in a way that attends to how artisans engage simultaneously in distinctive sorts of correspondence with the materials of their work as well as in distinctive relations of competition, collaboration, and conflict with other people. In addition to opening current contexts of artisanal production to insights from historical materialism and "new materialisms" (Coole & Frost, 2010) alike, such attention to artisanal assemblages seems especially appropriate given the preoccupations of artisans themselves. As ethnographic accounts attest, artisans' day-to-day lives are unquestionably influenced *both* by the vitality of the materials with which they correspond in their work *and* by the markets for what such correspondence produces. This recognition is especially true for miners classified as "artisanal," to whom I turn next.

Artisanal Mining as Artisanal Production

Observers often characterize artisanal mining in terms of how it *differs* from other sorts of resource extraction. Implicit in the definition of artisanal mining as "low-tech, labour-intensive mineral extraction and processing" (Hilson & McQuilken, 2014: 1), for example, is the notion that it is quite unlike the high-tech, mechanized mineral extraction and processing characteristic of large-scale industrial mining. In the following paragraphs, I take a different approach in framing artisanal mining by considering what it *shares* with other sorts of artisanal work.

Like the work of the artisans considered in the previous section, artisanal mining involves "small-scale, minimally capitalized commodity production and sales" (Colloredo-Mansfeld, 2002: 115) in the context of a global economy that simultaneously enables and marginalizes such efforts. As Lahiri-Dutt (2014) has recently described them, artisanal miners might be thought of as "extractive peasants," attempting to take "advantage, under considerable pressures, of the limited opportunities thrown up by their subordinate integration into new markets" (6). As with Herzfeld's cobblers, the pressures that render artisanal miners marginal are often most clearly manifested in more powerful actors' restricted and restricting visions of the value of their work. In the case of artisanal miners, however, it tends to be the chaotic, illicit, and inefficient, and not the "traditional," nature of their work that is stressed. As Lahiri-Dutt argues, many popular and scholarly accounts of artisanal mining offer justification for "mineral resource theories" that tend to "[exclude] and even [scorn] ... informal modes of mineral extraction in favour of institutions of mineral governance such as the state and the large corporations" (2014: 4; see also Hilson & McQuilken, 2014). In stressing the precarious nature of artisanal mining, however, such theories tend to overlook how such work is carried out by skilled practitioners who see potential where others may see only problems.

Like other sorts of artisanal production, artisanal mining is an activity in which practitioners make the most of "learned manual skills" (Herzfeld, 2004: 5) and use "tools ... that [extend] the mind/body into the environment" (Paxson, 2012: 121) – tools that, as Ingold might put it, mediate relationships of "correspondence" with the minerals they seek and the landscapes in which they seek them. Bryceson and Jønsson (2013) are especially clear on this point. Citing Sennett's influential reflections on "craftsmanship," they characterize the Tanzanian artisanal miners with whom they have done research as "autotelic craftsmen" (Bryceson & Jønsson, 2013: 17) – that is, as "self-propelled" (7) and skilled agents who are a far cry from the "economically irrational, socially destabilizing, and politically dissociated" (16) actors that some imagine them to be. "As craft producers," Bryceson and Jønsson write, artisanal miners "use their hands, heads and hearts, inadvertently fashioning a new local economy and society as they work" (8). In fact, research on the careers and social lives of artisanal miners reveals more stability, intentionality, and potential for sustainable development and communities than popular (and some scholarly) accounts of chaotic mining boomtowns might lead us to expect (Walsh, 2012; Peluso, 2018). Indeed, Bryceson and Fisher note how, like the artisanal producers described by Colloredo-Mansfeld, artisanal miners in Africa have carved out distinct niches of opportunity, and

> may be injecting a new sense of distributive justice and democracy. Rather than corrupting the social order, or continually being trounced by state and corporate mining interests, they have the potential to uplift their local communities and stimulate democratic principles ... [They are] breaking new ground, pioneering uncharted work lives, experimenting with novel

lifestyles and forming urbanizing communities in a frontier economy. (Bryceson & Fisher, 2013: 179–80)

Significantly, however, these "autotelic producers" (180) are pursuing all of these possibilities in the earlier described manner of artisans, meaning that their work lives, lifestyles, and communities are shaped by more than just their own hopes and plans. Their ambitions are realizable only to the extent that the materials, markets, and people with which they correspond in their work cooperate.

Like other sort of artisanal production, artisanal mining engages its practitioners in distinctive social relationships and political-economic systems that generate both precarity and potential. It also engages them in dynamic relationships with the materials of their work – not just the minerals they pursue and process, but the landscapes in which they find these minerals, the earth, stone, water, and other substances that enable or thwart their search, and the tools they use to make it all possible. As Peluso (2018) notes, small-scale mining territories are well understood as "emergent" and "entangled" territories, shaped by the "socialities of human labor, property relations, and knowledge practices" (414) in combination with the "materialities" of what is being mined. And what, exactly, is being mined matters a great deal. Although "[m]inerals do not have an essentialist character that ordains the outcome of mineral production and distribution" (201), there is no doubt that the wildly variable, non-anthropogenic, inherent properties of minerals demand different sorts of correspondence from the people who work with them.

Much of the precarity and potential associated with artisanal mining can be traced to what minerals themselves bring to this work. Consider, for example, Smith's reflections on how the inherent properties of coltan affect artisanal miners in the Eastern Democratic Republic of Congo, even though these miners know little of the digital uses to which this mineral will ultimately be put (2011: 18). Smith argues that, in being "relatively accessible" in the landscapes in which they are found, coltan and other digital minerals are well suited to artisanal extraction methods. What is more, in being "heavy and voluminous," these minerals necessitate a transportation infrastructure that requires teamwork and a broad division of labour, and enables taxation and state oversight (30). The specificity of these minerals' shaping of the "the social and political arrangement of [their] extraction" (21) is made even clearer through comparison with other minerals extracted in the same region. As Smith notes,

those who are knowledgeable about this business argue that the digital minerals, which all have similar material qualities, are better than precious metals and gems for building social connectivity and "movement" because they employ many people and give birth to a rich and dynamic division of labor ... Coltan, they say, does not enable people to become rich quickly, but it does allow them to profit consistently. (31)

As Smith implies, gold and gemstones engage people and enable social life differently than do digital minerals like coltan. Discussing small-scale gold-mining in West Kalimantan, Indonesia, Peluso also highlights gold's distinctive agency, describing not only how miners and traders understand it to be a particularly mobile, spiritual, and moody substance, but how, in concentrating value in a portable and liquid form, it enables (perhaps not so far-fetched) imaginings of how "shadow state actors at the high end of gold commodity chains" (2018: 411) secretively amass and move their wealth.

Considered agentive in these ways, minerals are revealed as more than passive tokens of anthropogenic value, stuff that is dug up and passed around a world wholly dominated by human intentions. As Smith puts it, minerals are "material substances [that] actively generate different kinds of social and political arrangements and entail different potentials for differently situated groups of people for profit or loss" (2011: 32). In the next section, I explore how this interpretation is valid in northern Madagascar with reference to the case of the region's artisanally mined sapphires.

Artisanal Sapphire Mining

Madagascar has long been known internationally as a rich source of mineral wealth. In one marketing campaign mounted by the Malagasy Ministry of Energy and Mines, the country is portrayed as a frontier destination in which international investors will find a wide variety of mineral resources awaiting exploitation. Not surprisingly, the island has hosted a good number of large-scale industrial mining projects over the years – from colonial-era gold and graphite mining ventures to recent projects intended to supply growing global demand for ilmenite, nickel, and other industrial minerals. Madagascar is also home to a large, and growing, artisanal mining sector that provides employment to an estimated 500,000 Malagasy people, some working full time at mining sites to which they have migrated internally and some mining only seasonally or part time to supplement other pursuits.

My research concerning artisanal mining in Madagascar has focused largely on the work of Malagasy miners and traders in the sapphire-rich region of Ankarana, 100 kilometres south of the provincial capital city of Antsiranana. In keeping with Smith's earlier noted reflections on how minerals themselves can be generative of social and political arrangements, I introduce this context in an unconventional way, for an anthropologist at least, by beginning, not with a historical or ethnographic account of the region's human occupation but, rather, with a gemological account of the specificities of the mineral around which so many of the people with whom I have been working have organized.

According to a gemological report published in *Gems and Gemology*, the journal of the Gemological Institute of America, a good portion of the corundum (better known to most as sapphires or rubies) found in the Ankarana region is classifiable as "blue-violet, blue, greenish blue, greenish yellow, and yellow (BGY) sapphires" (Schwarz, Kanis, & Schmetzer, 2000: 217). Deposited with other alluvial material throughout the region's limestone-riddled landscape over hundreds of thousands of years, these stones were first "discovered" (217) near the highway-side village of Ambondromifehy in 1996. Within a few years, this "deposit became one of the most productive sources of commercial-quality sapphire in the world" (217). One of the report's authors visiting the site in 1997 observed 10,000 Malagasy miners "active in the area," along with "Nigerian and Thai buyers [purchasing] much of the daily sapphire production" (217).

Although mechanized mining methods were used with some success in the early years of the boom, most of the sapphires coming out of the region originate, "unfortunately" (217), from within the boundaries of the Ankarana Special Reserve (now Ankarana National Park), "a nature reserve well known for its [limestone] karst topography" (217), which is accessible (albeit illegally) *only* to artisanal miners. In late 1998, the discovery of "extensive new alluvial deposits" (217) of sapphires and other gemstones further south in Madagascar drew many of Ambondromifehy's first wave of prospectors away. By 1999, however, many of these prospectors (and others) had returned. As of my own last visit in 2017, approximately 3,000 people remained in and around Ambondromifehy, most continuing to earn their livelihoods from mining and trading the region's sapphires (Walsh, 2012; 2015).

I first visited Ambondromifehy in 1997, early in the boom, and lived there over four months in 1999, by which time the region's previously non-existent sapphire mining and trading economy had become well established. In the years since, I have visited the town frequently, checking in with long-time acquaintances, asking after others who had left, and meeting new arrivals. Rather than summarize the findings of this research by stressing the experiences and perspectives of the miners and traders I have come to know, however, what follows takes direction from the gemological report cited earlier by attending primarily to how the materiality of sapphires and the landscapes in which they are sourced have influenced the people who correspond with them.

That Ankarana's sapphires are found in alluvial deposits means that they are accessible to people using relatively simple technology and techniques; most of the "diggers" referred to in passing in the gemological report use nothing more than mining bars, shovels, and flashlights in their work. Anecdotal reports suggest that, in the earliest months of the boom, sapphires were picked from the surface of the ground like pebbles from a basket of uncooked rice. Over time, however, the task of getting sapphires from the ground has become increasingly difficult and time consuming. Since 1999, certainly, all of the mining I have witnessed in the region has taken place in the "voids and crevices" (Schwarz et al., 2000: 219) of hard-to-access caves or at the ends of winding pits that have become deeper, longer, and more dangerous as they descend further into the alluvium. Over time, risk-taking and territory-making have become intertwined processes (Walsh, 2003), with the great value and relative inaccessibility of what miners are after continually heightening the peril and raising the stakes of their work, as in the case of Indonesian gold miners described by Peluso (2018: 410) and that of the Alpine crystal hunters discussed by Raveneau in chapter three.

Not surprisingly, such profoundly uncertain work has necessitated a variety of distinctive social arrangements. Miners excavating individual pits commonly work in teams, for example, often with the support of a patron or buyer who can feed them over the days, weeks, and, sometimes, months during which time they may come up with nothing. As with all social arrangements stemming from artisanal production, however, what has developed in Ankarana is not simply a product of the unchanging requirements of the materials with which miners correspond. That many prospectors worked illegally within a state-protected conservation area during the early years of the boom, for example, gave their mining a particular rhythm – they would arrive at mining sites under cover of darkness, work underground for as long as they dared, and then secret sacks of dirt through police patrols to water sources outside of the reserve for sieving. In the years since, circumstances and attendant arrangements have changed. The local police station was burned down in the midst of a nation-wide political crisis in 2003; in the years since, state conservation officials and police have played a limited role in controlling access to mining sites, allowing miners to settle inside the reserve for extended periods, working their pits more safely, regularly, and methodically than before. Following a pattern that Peluso describes in Indonesia, artisanal mining labour here *is* disciplined, but "the authoritative relationship is not located between miners ... and the state. It is between miners and each other, as well as between crews or individuals and [the mineral they are after]" (2018: 408).

Like Ankarana's miners, the region's gem traders must also negotiate the precarity and potential of work and social lives organized around sapphires. The business of making money from Ankarana's sapphires involves traders in something akin to what Geertz (1979) described as a "bazaar economy," that is, a context of interrelationship and exchange in which access to information about what is being traded can be as important as access to the trade-stuff itself. In such a context, miners and traders have developed considerable expertise in scrutinizing the "intimate antagonists" (225) with whom they bargain on a daily or weekly basis, learning as much as they can hope to know of what the other is looking for and what they are willing to pay for it. The sapphires over which buyers and sellers bargain play a role in this process as well. Since every sapphire is different, each must be evaluated as such, and not simply in terms of its colour, size, and internal features, but also how it instantiates a distinctive "bundling" (Keane, 2003) of these properties – a quality that sapphires share with other coloured gemstones, including emeralds (Brazeal, 2014; 2017). While Malagasy and foreign buyers' evaluations of individual stones always depend on what they know about the demands of other buyers to whom they will be selling, it is the inherent distinctiveness of the sapphires themselves that enables them to speculate in ways that they do, in ways that other precious minerals, like gold for example, would never allow (Walsh, 2004).

In certain ways, the sapphires mined in Ankarana are like sapphires from any other source in the world. For one thing, they are very durable, classified by mineralogists as second only to diamonds on the Mohs scale by which the relative hardness of minerals is measured. As finished gemstones, sapphires keep the shapes and polish they are given remarkably well over time and thus provide an ideal medium for the quality of timelessness so often attributed to them. Indeed, sapphires are often promoted by marketers as signifying a gift-giver's undying fidelity. Individual sapphires also tend to be quite small, an inherent property that makes them as easy to display in jewellery and other portable adornments as they are to conceal from any but those whose scrutiny is invited. In being both hard and small, sapphires are also able to concentrate the economic value attributed to them in a remarkably versatile form (Naylor, 2010). The vast majority of finished gem-quality sapphires circulating in the world today weigh less than three carats (600 mg); depending on its source and other factors, a single two-carat stone might sell to end consumers in many markets around the world for between US$500 and US$11,000. By even a conservative estimate, then, one kilogram of good quality two-carat sapphires, enough to fit into a pencil case, has a potential exchange value of US$1,000,000 – a fact that helps to explain both why sapphires are so easily and frequently smuggled internationally and why the Malagasy state has had such difficulty regulating their export (Duffy, 2007). Like the emeralds described in Brazeal's work (2014; 2017), and coloured gemstones more generally (Naylor, 2010), sapphires enable transnational, and potentially untraceable, movements of great wealth largely thanks to what they remain in the earliest stages of their social lives: relatively inconspicuous little coloured stones that might well pass for gravel.

Although the aesthetic qualities that consumers value in processed (that is, treated, cut, and polished) sapphires are not apparent to miners and traders in Ankarana, the material qualities of these stones nonetheless have a profound influence on how they engage people in the region. That sapphires are small, for example, means not simply that they are hard to locate, but that they are easily hidden when found; thus, a miner who finds a marketable stone at the end of a cooperatively worked pit might easily pop it into his mouth and sell it independently. In being so small, sapphires are also things that independent, mobile traders can buy, store, and travel with in anticipation of an eventual sale to a foreign buyer who may come to the region only once a week. The great durability of Ankarana's sapphires, meanwhile, is among the qualities that most distinguishes them from other commodities produced or procured locally for distant, and largely unknown, markets. While shark fins and sea cucumbers, for example, begin deteriorating once they become marketable, and thus must be sold quickly at whatever price is being offered by the intermediaries who will see to their export, sapphires can be held on to for months or even years after extraction in anticipation of a potentially profitable future sale. Indeed, what is perhaps most distinctive about sapphires as resources in Ankarana is how they have enabled local traders to speculate *in* the commodity with which they have become engaged in ways that are usually reserved for others further along global commodity chains. Still, there are significant limits to what local traders (can) know of what will happen to the sapphires they will eventually sell to others – mostly Thai buyers – who take them out of Madagascar.

The purpose of the previously cited gemological report is not just to describe Ankarana's sapphires, but to describe their *potential* to readers with interests and expertise of their own. Thus, the report attends to these stones' "visual appearance" (Schwarz et al., 2000: 222), their "crystal morphology" (223), their "microscopic" (224) and "spectroscopic properties" (227), and their "chemical composition" – inherent features that combine to give all sapphires the colours, shapes, and sparkle (among other qualities) that consumers expect to find in finished gemstones. Just as relevant is the report's attention to the susceptibility of Ankarana's stones to "heat treatment" (222), a process through which cloudy and plain-looking rough sapphires might be transformed into clearer, more brightly coloured, and, hence, more valuable finished gems. Not discussed in this report are other, less well-publicized, enhancement techniques that Ankarana's sapphires might secretly undergo in Thailand's processing labs – coating stones with beryllium, for example, before heating them in order to effect even greater transformations of appearance (Hughes, 2002). Still, the report offers a glimpse of how the potential of Ankarana's stones is reckoned by some, not simply in terms of what they are straight out of the ground but in terms of what they might become through careful processing undertaken by skilled specialists outside of Madagascar.

The report concludes with a discussion of how gemologists will be able to discern Ankarana's sapphires from the "synthetic" or "laboratory-grown sapphires" (Schwarz et al., 2000: 231) with which some might unwittingly confuse them – a task that should be "fairly simple" to an "experienced gemologist" following "careful microscopic examination" (231). That the possibility of confusing sapphires sourced in Ankarana with those sourced in a laboratory even exists, however, suggests that Ankarana's sapphires have influenced the development of social life in this region, not only by virtue of what they are materially but also by virtue of how they have come to be. Although lab-produced sapphires are materially identical in most ways to the most highly valued of mined sapphires, such synthetic stones are not valued nearly as highly by gem lovers and jewellery consumers as are their natural counterparts (Walsh, 2010). The fact that the "natural" origins of sapphires matter to many gemstone consumers clearly matters a great deal to people in Ankarana. Simply put, if their origins didn't matter, Ankarana's sapphires would today remain what they were prior to their "discovery" in 1996: little bluish, greenish, yellowish stones with no potential.

In approaching artisanal sapphire mining in Ankarana in the round-about way I have here, I hope to have illustrated two points: first, how the inherent properties of sapphires and the landscapes in which they are found figure in shaping the human-mineral engagements essential to local and international sapphire trades; and, second, how artisanal mining, like all artisanal production, engages people in

complex assemblages involving not just minerals and exchange partners but also landscapes, tools, policies of resource governance, and
so on. The intent has been to draw attention to the complexity and
vitality of a category of work – artisanal mining – that is too often
dismissed as "crude" and "backward." What remains to be considered is how a focus on the artisanal side of artisanal mining might
reveal something more than just the complexity of human-mineral
engagements. In the following, concluding, section of this chapter,
I suggest one possibility.

Outcrops of the Anthropocene

In a lecture delivered at the 2014 annual meeting of the American
Anthropological Association, Bruno Latour offered anthropologists a
"nudge" intended to move us "forward," towards a fantastic gift that
is ours for the taking: the gift of the "Anthropocene," the term that
geoscientists commonly use to describe our current geological epoch
(Latour, 2014). The gift's wrapping leaves little to the imagination, however, and the central question that anthropologists faced with this gift
must answer is not, "What could it possibly be?" Rather, the question
we must ask ourselves is, "What are we to do with this gift?" To conclude, I propose that one way of approaching the Anthropocene as an
anthropologist is through the careful study of sites and processes, like
the ones described in the preceding pages, in which humans and minerals share in artisanal correspondence.

While clearly anthropologists and geoscientists have much to learn
from one another, we also have more in common than some might
assume. Here, I write from experience, as my father-in-law, Fried
Schwerdtner, is a structural geologist with whom I connect on many
levels, but, most obviously, over something we have always shared:
fieldwork. For Fried, fieldwork has meant more than forty years of wandering Canadian landscapes in search of "outcrops" – sites at which the
earth's deep history sits exposed above the surface and, thus, sites at
which the effects of key processes of past geological epochs are readily
apparent and open to analysis. While I may not see what he does when
he looks at an exposed cliff or a road cut, I certainly recognize something familiar in what guides and keeps his attention to and at such
sites. In approaching outcrops in the ways they do, geologists engaged
in fieldwork are able to infer complex stories from the most mundane
stuff imaginable. Like anthropologists who avow the value of deep
considerations of what others take for granted, geologists generate

insights from searching out and thinking through the observable traces of complex processes at work.

Bearing in mind that much of what we know of the geological time-line that runs into the Anthropocene has been learned from research on outcrops, it stands to reason that humans are currently engaged in producing what may become our species' most enduring monuments – the outcrops by which a condensed version of the current epoch's story might be told hundreds of thousands, or millions, of years from now. But why wait? Why not try to tell a more expansive story of the Anthropocene now? And why not start by identifying outcrops of our own – sites at which the effects of the characteristic processes of the *current* geological epoch are especially apparent and open to analysis, and sites to which we might point others in trying to educate them? What sites might these be? Taking direction from Latour, we might envision sites that are especially conducive to helping us think beyond familiar dualisms of culture and nature, social and material, anthropological and geological, and so on. Or, in line with Chakrabarty's influential work on the topic, we might envision outcrops of the Anthropocene as sites at which "global histories of Capital" might be found "in conversation with the species history of humans" (2009: 212). The more speculative envisioning we do, however, the clearer it becomes that we, like geologists, might find outcrops just about anywhere.

One obvious reason for considering Ankarana's (and the world's) artisanal mining sites as outcrops of the Anthropocene is because they are sites at which people are systematically altering the earth beneath their feet. However much some imagine these sites as anarchic or chaotic, the mining that goes on at them unquestionably leaves distinctive traces revealing processes of extraction and "territorialization" (Peluso, 2018). What these traces (will) tell is stories of engagement patterned not simply by the whims of humans but by the underground "paths" (as miners put it) already established by the minerals they pursue and extract. In connecting these outcrops to others, future geologists might also infer broader, global processes that have enabled the widespread shifting of our planet's most durable matter. Just imagine the story that the irregular global distribution of sapphires mined in Madagascar might tell millions of years from now: stories not just of the reach of the twenty-first century sapphire trade, but of the epoch of which this trade is just a small, though maybe symptomatic, part. Again, though, why wait?

I am not suggesting we look to artisanal mining sites for people to blame for the problems associated with the Anthropocene – these are not places in which to find "highly localised networks of ... individual bodies whose responsibility [for the ills of the Anthropocene] are staggering" (Latour, 2014). Research at these sites does, however, offer opportunities for considering how the spreading, gutting edge of capitalism is now, as it has done for some time, entangling humans *with* the earth they are remaking in distinctive and mutually defining ways.

As discussed previously in this chapter, a focus on the artisanal nature of artisanal mining draws attention to how miners and petty traders like those with whom I have been doing research in Madagascar are involved in complex assemblages through which they engage, simultaneously, with other people, with local and global mineral markets, and with the very matter of the landscapes in which they work. As artisans in both of the senses noted earlier in this chapter – that is, as people who work in ways that exemplify how *all* humans correspond with the matter of the world around them and as people engaging with global markets in anything but anachronistic ways – artisanal miners represent the *anthropos* of the Anthropocene well, in that their work requires us to think beyond familiar dualisms even as it presents us with a particular segment of "the global history of capital" in the making. Put another way, Ankarana's artisanal miners' distinctive engagements, like those of all artisans, might be understood as especially revealing of ways of being and becoming that are fundamental to our species and the geological era we have wrought. In the manner of basket weaving, brick making, and other exemplary forms of artisanal work, artisanal mining lays bare how it is that humans have managed to implicate ourselves in what Ingold terms a "process of correspondence" with the non-human world through which we have been "drawing out or bringing forth potentials immanent in a world of becoming" (2013: 31) – a world that in becoming what we have been and are bringing forth in it may well soon be done with us. Further research focusing not just on artisanal mining but on human-mineral engagements more broadly offers important opportunities for teasing out the complexity of this ongoing process.

Acknowledgments

The research on which this chapter is based was funded by the Social Sciences and Humanities Research Council of Canada. I am grateful to the people of Ambondromifehy and surrounding communities for all they have taught me and to many others whose work and comments have helped to shape what's presented here, especially Lindsay Bell, Josh Bell, Brian Brazeal, Filipe Calvão, Elizabeth Ferry, Les Field, Susan Gillespie, Richard Hughes, Gilles Raveneau, Tania Richardson, Fried Schwerdtner, Annabel Vallard, and Gisa Weszkalnys.

REFERENCES

Antrosio, J., & Colloredo-Mansfeld, R. (2014). Risk-seeking peasants, excessive artisans: Speculation in the Northern Andes. *Economic Anthropology, 1*(1), 124–38. https://doi.org/10.1002/sea2.12008

Bennett, J. (2010). *Vibrant matter: A political ecology of things*. Durham, NC: Duke University Press.

Brazeal, B. (2014). The fetish and the stone: A moral economy of charlatans and thieves. In P.C. Johnson (Ed.), *Spirited things: The work of "possession" in Afro-Atlantic religions* (pp. 131–54). Chicago: University of Chicago Press.

Brazeal, B. (2017). Austerity, luxury and uncertainty in the Indian emerald trade. *Journal of Material Culture, 22*(4), 437–52. https://doi.org/10.1177/1359183517715809

Bryceson, D.F, & Fisher, E. (2013). Artisanal mining's democratizing directions and deviations. In D.F Bryceson, E. Fisher, J.B. Jønsson, & R. Mwaipopo (Eds.), *Mining and social transformation in Africa: Mineralizing and democratizing trends in artisanal production* (pp. 179–206). London: Routledge.

Bryceson, D.F, & Jønsson, J.B. (2013). Mineralizing Africa and artisanal mining's democratizing influence. In D.F Bryceson, E. Fisher, J.B. Jønsson, & R. Mwaipopo (Eds.), *Mining and social transformation in Africa: Mineralizing and democratizing trends in artisanal production* (pp. 1–22). London: Routledge.

Chakrabarty, D. (2009). The climate of history: Four theses. *Critical Inquiry, 35*(2), 197–222. https://doi.org/10.1086/596640

Colloredo-Mansfeld, R. (2002). An ethnography of neoliberalism: Understanding competition in artisan economies 1. *Current Anthropology, 43*(1), 113–37. https://doi.org/10.1086/324129

Coole, D., & Frost, S. (Eds.). (2010). *New materialisms: Ontology, agency, and politics*. Durham, NC: Duke University Press.

Dorner, U., Franken, G., Liedtke, M., & Sievers, H. (2012). *Artisanal and small-scale mining (ASM)*. POLINARES working paper no. 19. Retrieved from https://www.polinares.eu/wp-content/uploads/2017/12/polinares_wp2_chapter7.pdf

Duffy, R. (2007). Gemstone mining in Madagascar: Transnational networks, criminalisation and global integration. *The Journal of Modern African Studies, 45*(2), 185–206. https://doi.org/10.1017/s0022278x07002509

Garcia Canclini, N. (1995). *Hybrid cultures: Strategies for entering and leaving modernity*. Minneapolis: University of Minnesota Press.

Geertz, C. (1979). Suq: The bazaar economy in Sefrou. In C. Geertz, H. Geertz, & L. Rosen (Eds.), *Meaning and order in Moroccan society: Three essays in cultural analysis* (pp. 123–314). Cambridge: Cambridge University Press.

Giordano, C. (2002). Commentary on "An ethnography of neoliberalism: Understanding competition in artisan economies." *Current Anthropology* 43(1), 127–8.

Hentschel, T., Hruschka, F., & Priester, M. (2003). *Artisanal and small-scale mining: Challenges and opportunities*. London: International Institute for Environment and Development.

Herzfeld, M. (2004). *The body impolitic: Artisans and artifice in the global hierarchy of value*. Chicago: University of Chicago Press.

Hilson, G., & McQuilken, J. (2014). Four decades of support for artisanal and small-scale mining in sub-Saharan Africa: A critical review. *The Extractive Industries and Society*, *1*(1), 104–18. https://doi.org/10.1016/j.exis.2014.01.002

Hughes, R. (2002). The skin game. Retrieved from http://www.palagems.com/treated-orange-sapphire

Ingold, T. (2007). Materials against materiality. *Archaeological Dialogues*, *14*(1), 1–16. https://doi.org/10.1017/s1380203807002127

Ingold, T. (2012). Toward an ecology of materials. *Annual Review of Anthropology*, *41*(1), 427–42. https://doi.org/10.1146/annurev-anthro-081309-145920

Ingold, T. (2013). *Making: Anthropology, archaeology, art and architecture*. London: Routledge.

Keane, W. (2003). Semiotics and the social analysis of material things. *Language & Communication*, *23*(3–4), 409–25. https://doi.org/10.1016/s02715309(03)00010-7

Lahiri-Dutt, K. (2014). *Extracting peasants from the fields: Rushing for a livelihood?* Singapore: Asia Research Institute, National University of Singapore.

Latour, B. (2014). Anthropology at the time of the Anthropocene – A personal view of what is to be studied. Distinguished lecture delivered at the annual meeting of the American Anthropological Association, Washington, DC. Retrieved from http://www.bruno-latour.fr/sites/default/files/139-AAA-Washington.pdf

Naylor, T. (2010). The underworld of gemstones. *Crime, Law and Social Change*, *53*(2), 131–58. https://doi.org/10.1007/s10611-009-9223-z

Paxson, H. (2012). *The life of cheese: Crafting food and value in America*. Berkeley: University of California Press.

Peluso, N.L. (2018). Entangled territories in small-scale gold mining frontiers: Labor practices, property, and secrets in Indonesian gold country. *World Development*, *101*, 400–16. https://doi.org/10.1016/j.worlddev.2016.11.003

Schwarz, D., Kanis, J., & Schmetzer, K. (2000). Sapphires from Antsiranana province, northern Madagascar. *Gems & Gemology*, *36*(3), 216–33. https://doi.org/10.5741/gems.36.3.216

Scrase, T.J. (2003). Precarious production: Globalisation and artisan labour in the Third World. *Third World Quarterly*, *24*(3), 449–61. https://doi.org/10.1080/0143659032000084401

Smith, J.H. (2011). Tantalus in the digital age: Coltan ore, temporal dispossession, and "movement" in the Eastern Democratic Republic of the Congo. *American Ethnologist*, *38*(1), 17–35. https://doi.org/10.1111/j.1548-1425.2010.01289.x

Walsh, A. (2003). "Hot money" and daring consumption in a Northern Malagasy mining town. *American Ethnologist*, *30*(2), 290–305. https://doi.org/10.1525/ae.2003.30.2.290

Walsh, A. (2004). In the wake of things: Speculating in and about sapphires in Northern Madagascar. *American Anthropologist*, *106*(2), 225–37. https://doi.org/10.1525/aa.2004.106.2.225

Walsh, A. (2010). The commodification of fetishes: Telling the difference between natural and synthetic sapphires. *American Ethnologist*, *37*(1), 98–114. https://doi.org/10.1111/j.1548-1425.2010.01244.x

Walsh, A. (2012). After the rush: Living with uncertainty in a Malagasy mining town. *Africa*, *82*(2), 235–51. https://doi.org/10.1017/s0001972012000034

Walsh, A. (2015). Lost and/or left behind in a Malagasy outpost of the "underworld of gemstones." *Critique of Anthropology*, *35*(1), 30–46. https://doi.org/10.1177/0308275x14557094

York, G. (2012, 18 August). Young and dying: The scandal of artisanal mining. *The Globe and Mail*. Retrieved on 22 October 2013 from https://www.theglobeandmail.com/news/world/young-and-dying-the-scandal-of-artisanal-mining/article4487572/

3 The Value and Social Lives of Alpine Crystals

GILLES RAVENEAU

This chapter considers the organization and effects of the circulation of Alpine minerals among different social fields (amateur, professional, scientific, and patrimonial), focusing especially on the processes of commodification and de-commodification that characterize this circulation. My approach builds on the idea that crystals, like people, have biographical trajectories, which lead them through multiple "regimes of value" (Appadurai, 1986) over time, and that much can be gained from following these trajectories and considering how they are (or are not) valued and evaluated as they proceed from one actor and/or place to another (Appadurai, 1986; Bonnot, 2002; Latour, 1994; Kopytoff, 1986).

What follows is based on long-term ethnographic fieldwork (2002–2012) in France, Switzerland, and Italy. This research was originally aimed at understanding the logic of relationships with space in high mountain areas, its exploitation, and the organization of crystal hunters in certain territories that shape social inclusion hinged on risk and risk-taking. Over the years, the project has developed to include a focus on what happens to crystals when they are appropriated by others (dealers, private collectors, museums, and so on) in various sociocultural contexts, requiring me to trace their paths from sites of hunting and discovery in the mountains to their destinations in collections and markets.

This survey was conducted in a spirit of commitment and shared experience in the mountains with some crystal hunters with whom I was able to build a reliable bond over time. As a mountaineer, I was able to gradually accompany them on their crystal routes and understand the living conditions of autonomy in high altitude mountain bivouacs as well as the use of shelters, the search for crystals on cliff faces, their discovery, the working of "ovens,"[1] the transport of crystals to the valley, and their subsequent travels. This experience shared with crystal hunters revealed conflicting and agonistic relations between individuals from the same mountain range. If penetrating the world of crystal hunters was long and difficult, research with collectors was also complicated. Collectors are always anxious to protect their collections from curious observers such as tax authorities, especially in a context where some had their collections confiscated following a lengthy trial (2006–2008) in which crystal hunters and collectors were accused of asset concealment. The distrust and mistrust are simultaneously so widespread that it makes the work of the ethnologist who wants to go beyond false appearances – to understand what is going on in the search for crystals as well as in their exchanges and their sale afterwards – fascinating and very instructive.

This chapter addresses the exchanges involved in the movement of quartz and other minerals and the changes that have accompanied such movement. I focus on how these crystals and minerals change status when they pass from one actor and/or place to another, becoming goods at one point in their existence and ceasing to be so subsequently, and vice versa. My objective is to show how the value of these minerals emerges from the exchanges they undergo, and how their economic value is the product of social exchange and not the reverse. From this point of view, the value and prestige of minerals are not only in their objective materiality but also result from the exchanges and symbolic production attributed to them by their users and apparent in their biographical itineraries. Not only do they have an economic or patrimonial value, but they are also the carriers of social distinction and singularity. Before examining the attribution of value to Alpine minerals, however, it is necessary to establish what the activity of crystal hunters is.

Treasure "Ovens" and Mountains

Crystal hunters roam the Alps in search of minerals and quartz crystals. The modes of crystal hunters' experiences are different from those of climbers. Mountaineers aim at sporting prowess or the simple pleasure of climbing. They seek to reach a summit, following a known path via the consultation of a guidebook or, more exceptionally, for the best of them, to open up a new, never-before-used trail for ascending a mountain. In all cases, they avoid crumbly walls and "rotten" rocks. Crystal hunters seek just the opposite. They explore the unusual paths, places, and crumbled walls that others avoid, roaming over mountains with the (virtually) sole goal of searching for crystals. Indeed, it is often on altered grounds, where walls disintegrate and uncover cracks, that crystal hunters discover the crystal-rich "ovens" they are seeking. This experience of randomness and risk-taking plays a key role in how crystal hunters deal with the environment. Immersed in the high mountains, as they are several days on end, crystal hunters engage with an Alpine environment that determines the space, time, and forms of action that enable them to remove the valuable minerals they seek. By choosing the logic of discovery and a direct confrontation with nature, crystal hunters embody a specific social inscription based on risk and risk-taking, as fatal accidents testify. Yet, this work is also closely related to a treasure hunt. Crystal hunters pursue the quest for a virgin "oven," the miraculous crack full of "mature" crystals that have been waiting for them for millions of years, which they will be able to take down intact and pristine. They are all looking for this treasure, buried in the hollow walls, hidden at the tops of mountains – a treasure that can be exchanged for money quickly and easily, albeit only for those who are willing and know how to take risks and jeopardize their physical integrity.

The financial gains promised by crystal hunting are an important element of this activity. The profusion of mineral exhibitions and mineral markets reflects a public interest in these items. People exchange and sell minerals of various sizes and value. Prices range from a few dollars to thousands, even tens of thousands, of dollars or more for exceptional pieces that kindle the imagination of collectors. The finest specimens are rarely shown in mineral shows. The great collectors often visit the crystal hunters directly to try and negotiate their discovery before it is exhibited and made public. In France alone, approximately 250 exhibitions of minerals took place in 2014. These exhibitions range from small local events to gatherings of international renown, such as the one that takes place at Sainte-Marie-aux-Mines.

It is also important to acknowledge the profound meaning with which this work is loaded, especially in its association with the presence of precious and rare materials: rock crystal and other types of unusual minerals, such as fluorite, amethyst, axinite, epidote, or siderite, with or without quartz. Alpine crystals, more particularly those of the Mont Blanc massif, are generally specific to the quartz vein cavities. They come in different forms (comb, gwindel, sugar lump, spike, more or less elongated, scepter, bipyramid) and different colours (smoky, clear, with chlorite deposit on the surface), loose in a group or attached to the rock, and are combined or not with other minerals such as fluorite, amethyst, axinite, or siderite. It has to be noted that rarity is a relative concept, which depends not only on the natural content of the materials but also, and especially, on many political, social, and symbolic factors. Rarity, preciousness, and value establish a complex system that largely deviates from the natural order of things.

Professional crystal hunters who work full time at the task are few. The vast majority of crystal hunters working in the Alps carry out this work as a side activity, partly because crystal hunting is seasonal work, tied to weather conditions, and generally carried out from June to September. But the main reason most people only hunt crystals part time is because the search for minerals is very uncertain, and the income it generates is directly related to the quality of the crystals found and thus fluctuates considerably, depending on a hunter's discoveries or non-discoveries. Occasionally, exceptional discoveries (like the beautiful red fluorites found in 2008 on Mont Blanc) ignite the imagination and arouse vocations. But such discoveries are very rare, and the financial gains they promise are hard to achieve. Only long practice, a good knowledge of the natural environment, and skilled mountaineering make it possible to discover the treasure that so many pursue.

The crystal hunter may be an engineer, a craftsperson, a mountain guide, a company employee, or a professional person; most commonly, they are people with jobs that allow them the freedom to manage their own time. Indeed, what differentiates crystal hunters from simple mineral collectors is not only the quality and the quantity of specimens they discover, but also the energy, the means, and the time spent looking for minerals in the mountains. Another factor that sets crystal hunters apart from the crowd of amateurs is that they often seek to trade or sell, rather than keep, their rock crystal discoveries, even though many of them are collectors themselves. Finally, what characterizes all of the crystal hunters with whom I have conducted research – like the scrappers described by Bell in chapter one and the prehistoric bead collectors in Thailand discussed by Vallard in chapter four – is the passion they have for this activity, the mountains, and a lifestyle based on freedom, wilderness, discovery, dreams, adventure, and risk. Their passion for this activity, its connection with death, and the material involved are all associated with a quest for treasure (Raveneau, 2006). According to the usual terms, rock crystal is "the gold of crystal hunters" (Canac, 1980). To get it, however, is a way of making money that involves imagining the discovery of a treasure – the unique, the virgin "oven" – while betting that one's life will not be taken in the process. Of course, crystal hunters' own standing among peers and others is also in the balance and depends on their ability to remove these precious minerals from the mountains in which they find them.

Compared with the risks taken in the mountains, the issues and risks involved in crystal hunters' social relationships may seem trivial. Men engage in rivalry or cooperation with one another, but their lives are not threatened by all that. Yet, something fundamental is at stake in such engagements and interactions: social status. Crystal hunters negotiate a shared social world in which rivalries, competition, pride, honour, death, lies, accidents, passion, luck, and money are ever-present, and they organize themselves according to the results of the interactions entailed by the workings of this world. The group establishes itself through their sometimes emotionally charged relationships with one another. Agonistic exchanges create the community field of crystal hunters and define the group autonomy and the field of prestige they share.

Agonistic Exchange and the Value of Crystals

Conflicts over territory and access to "ovens," conflicts that emerge in the trade and sale of crystals, and the internal conflicts emerging in teams of crystal hunters are always linked in one way or another to particular minerals, to their value, and to the financial benefits they might produce if sold. In other words, money is always a factor worth considering in trying to understand these conflicts. It is important, however, to consider other factors as well. Prestige is another essential thread to consider. When a crystal hunter wants a complete victory, it is not so much the financial gain that is being sought, but the symbolic gains of prestige and fame. This question clearly appears in the story of the discovery of the crystal called "Laurent."

In the summer of 2005, Christophe, a Parisian crystal hunter I knew who had spent summers in Chamonix for thirty years, joined his friend Laurent, a mountain guide, for a few crystal runs in the Rochassiers, an area in the Mont Blanc massif. During one of their outings, while they were rappelling on a rock wall, Laurent fell to his death. Christophe was very affected by the accident. I met him a few days after the accident when he showed his discoveries at the Chamonix mineral market, which was held on the first weekend of August and brought together crystals, crystal lovers, collectors, enthusiasts, and buyers. The next year, in July 2006, despite a bad season and falling rocks, Christophe discovered an "oven" in the Aiguilles Vertes of the Mont Blanc massif from which he extracted fluorites of various colours and different sizes, ranging from purple to dark red. He worked there for days. On 21 July, after more than five hours of difficult and careful work, while the rocks around him threatened to collapse, he extracted an incredible specimen: a rare association of two minerals typical of the Chamonix Mountains, red fluorite and smoky quartz. The whole specimen was of rare beauty: six centimetre octahedrons of an intense red fluorite arranged like a lava flow on the tip of a twenty centimetre smoky quartz. Christophe understood immediately that he had just made an exceptional discovery, the one he had been anticipating for more than thirty years, the dream of every crystal hunter. This feeling was confirmed in the following hours by fellow crystal hunters who came to see the spectacular mineral upon hearing news of its discovery. Very quickly, Christophe decided to call it "Laurent," as a tribute to his rope partner who had died the previous year.

I was on site in July 2006 when Laurent was discovered. I made several trips there that summer with Christophe, but, unfortunately, I was up the mountain with another crystal hunter on the day it was found. On subsequent days, however, I was able to attend the procession of crystal hunters, collectors, cultural mediators, mineralogy enthusiasts, and others, and witness the inflamed arguments, attitudes, and emotions produced by this revelation. Within the week following the discovery, the great collectors likely to be interested were already competing to acquire it. I subsequently regularly followed the different phases, controversies, and proposals of purchase through which Laurent proceeded, until its acquisition by the Museum of Natural History in Paris for the sum of €250,000 and its classification as "cultural property of major patrimonial interest," a supreme recognition for the discoverer and the crystal since it is the first mineral to be classified as such.

Although the money received upon selling a particular specimen helps to establish one's status, indicating a hunter's ability to find and bring back precious minerals and providing an objective and legible measure of material success in everyone's eyes, it may nevertheless be spent as quickly as it has been earned, "burnt through" in a few days, often by the youngest and most inexperienced crystal hunters. The mountain appears to be a permanent and infinite source of wealth to those who know it and are willing to take the necessary risks.

Anthropological literature attests well to the sometimes ambiguous status of easy money earned through risky mining work. For example, Pascale Absi (2003) describes how the miners of Potosí, Bolivia, associate money earned from mining with the devil, and treat it accordingly. Similarly, Andrew Walsh (2003) describes how young male sapphire miners in northern Madagascar spend a great deal of what some call "hot money" to satisfy their immediate desires rather than investing in long-term plans. For Alpine crystal hunters, money alone is not a sufficient motivator, meaning that "those who do it for money" are discredited by others who invoke other criteria in describing their motivations. Those who seem to "do it for money" are mocked not only for their greed but for their stupidity, looked down upon by others, generally the oldest in the business, who claim to understand what this work is really about. These long-time hunters have made major discoveries, and sometimes a great deal of money as well, but they also have had disastrous seasons and many mountain runs that yielded nothing. Money or not, they continue to work, faithful to their passion for minerals.

Put another way, the money that hunters earn from selling crystals is not as important as what circulates with, and is signalled by, that money: the meaning that runs through it all. Financial gains are as much the measure of usefulness as a symbol of moral importance. It is the prestige and reputation associated with the discoveries leading to sales that make the risks involved in this work worthwhile. Hunters not only risk life and death in the mountains; they face metaphorical death as well, because, while losing face or being humiliated may not literally kill you, it may do so symbolically.

Although crystal hunters do not explicitly formulate it in this way, it may not be so much the money that counts in these conflicts and rivalries as the value of crystals, that is to say, not only their monetary value, their sale price on the economic market, but also their social value in terms of the recognition, esteem, dignity, manhood, honour, virtuosity, power, and other relevant concepts that crystallize in hunters' relationships with one another. Rivalry is produced and organized through the pursuit of recognition and not simply profit.

This drive for recognition does not mean that money does not matter. On the contrary, it would be absurd to think that the material gains to be earned from this activity have no significance. Crystal hunters are willing to risk a lot to find and extract precious materials from mountain ranges, and some of those who are determined to sell part of their discoveries earn a portion of their living from this work. Thus, they cannot take lightly the income they expect to earn from crystal hunting, and they appreciate how such earnings contribute indirectly to the establishment of their status and to the balance of power among peers. One could say that, although money is not the main or sole consideration for crystal hunters, it provides a stamp of legible value that unquestionably brings prestige.

What gives looking for crystals its dramatic intensity is, of course, the risk of death, which is the ultimate challenge; but this is reckoned to be between a crystal hunter and the natural elements, meaning that the individual faces it alone. Risk-taking and death are also at the heart of the collective identity of this group – both are commonly referenced in the appreciation of the personal value of each hunter. As soon as a crystal hunter gets back to the refuge or to the valley, he finds himself with others; then, what gives the activity its value is not the money itself, represented by minerals and what they can sell for, but what takes place because of the exchange value of crystals: the hierarchy of social positions in the group and the tensions that establish interactions for differentiation and prestige. The price of crystals brings up the conflict of values and the agonistic element that underpins that conflict.

The kind of death involved in the structure of social relations is neither real nor imaginary. It establishes a price, a symbolic guarantee of the individual's value. It is used between men to establish everyone's relative value in the light of this commitment. In other words, it is the assessment by others of the genuineness of the subject's commitment to his action. This assessment cannot be done without the intervention of death (as a symbolic function) and without the repression of the fear of death as a sign of personal value. It is death that the subject throws in the face of others in response to any doubt they may have with regard to his value (Geffray, 2001). Death, the ultimate signifier, backed up by the find of crystals, achieves the establishment of faith in the value of all as members of the group. Prestige and honour are products of the collective consent given to the value of the physical commitment to the activity, attested to by the confrontation with death (Raveneau, 2006).

We know that, for the discovery of the crystals, men are willing to risk the greatest stakes, combining vertigo with loss of control of the situation. However, what is problematic here is not so much the playing with euphoria and vertigo, but the way this desire is focused on minerals. What precisely are the relationships between crystals, death, and money? How do they interact?

Anthropology and the history of religions place the origin of money in the successive substitutions for human sacrifice. We can see in the series of substitutions (animals, grains, and so on) a reduction in the cost of this exchange. But, the nature of its original purpose – the sacrificial nature of the transaction – remains paramount. George Simmel (1987), in his book *Philosophy of Money*, draws on this principle to dispute a key tenet of classical economics by saying that it is the price, that is to say the sacrifice, that determines the value, and not the value that determines the price.

Building on this idea of sacrifice, I propose to explain why it is that crystal hunters are willing to risk their own lives to bring back the precious crystals they seek. These crystals are precious because they are rare and because they come from the "bowels" of the wilderness, yes, but they are also precious because human lives are sacrificed to obtain them. To be ready to risk one's life, daring to transgress the supreme prohibition (that of putting one's life in danger, of looking death in the eye) seems both to heighten the status of those who do it and to add value to the minerals they bring back.

For the ancient Romans, the best prey was that which cost the lives of its hunters. Pliny notes that death distinguishes "good" wild game. In the same way, the death associated with obtaining an object can make it precious. From this point of view, death can generate the value we attach to certain goods. That men risk their lives to remove rock crystals from the mountains is therefore essential to the value attached to these minerals. But this value goes deeper still when we consider the threat of death at stake in the mineral's acquisition alongside the magical virtues lent to minerals used in the traditional healing practices of the Newar of Nepal (Manandhar, 1998), or the Tibetans (Meyer, 1988), or even in the modern crystal therapy of Western societies. Besides the value of the minerals themselves as precious goods, there is also the value of the men who get them. Ultimately, trust and recognition exist only if each one is fully engaged in the activity and accepts the risk to life. All this value is concentrated in the harvesting of crystals. From this point of view, crystals are objects of belief. They establish a principle of identity in which crystal hunters invest their faith. The symbolic dimension of the group, partly built by its members, is inseparable from the constitution of the group itself and from the various forms the feeling of belonging to the group can take for the members who make it up. In the same way that shells, elsewhere, are exchanged against women or to compensate for the death of a warrior (Malinowski, 1989; Panoff, 1980; Godelier, 1982), crystals stand, so to speak, as symbolic substitutes for human beings, as imaginary equivalents of life, power, and wealth.

Just like Laurent, many exceptional minerals bear a name (Amédée, Georges, Daniela, Pinelle, Hoppe, Cullinan, and so on) taken from their discoverers or someone else associated with the specimen's history or the person who discovered it. We are here in a regime of singularity. I would like to complete my hypothesis – that crystals are objects of belief that constitute a principle of identity – and go further by considering the question of the price and the value of crystals, their use and exchange value on an economic market where they are sold and exchanged, and by linking this issue to the concept of "prestige goods."

Merchandise or Prestige Goods?

Crystals are the object of various transactions between crystal hunters, between crystal hunters and collectors, and among collectors themselves. The minerals move around as gifts, through bartering, and in commodity and non-commodity exchanges. A non-commodity exchange is an exchange in which social relations prevail, that is to say, "an exchange conditioned by another social relationship which goes beyond it, both because it commands it and generally because it survives it" (Testart, 2007: 143; see also Gregory, 1982; Leach & Leach, 1983). In a commodity exchange, on the contrary, the relationship between people is based only on the consideration of the price of the good. We must acknowledge that crystals are sold and purchased. They have or may have a price, and, as such, crystals are merchandise. But are they really goods like any others?

At this point of reasoning, it should be noted that some crystals are neither given nor bartered for others, or even sold, but kept by hunters for themselves, held incommunicado in a collection. They are not exchanged any further, and they enter the field of "inalienable possessions" (Weiner, 1985: 210–27). Keeping something for oneself is the same as giving it to someone else. We give and we sell also because we keep or because we can keep better by this roundabout way. This point, then, changes the perspective on donated and sold crystals, and on the price and the value of these precious objects.

Taking crystals seriously in their singular reality and their complexity is acknowledging that they have some power (Bazin & Bensa, 1994; Bazin, 1997) and admitting that they belong to another order of reality than objects that are simply exchanged by virtue of the sacrosanct principle of reciprocity (Lévi-Strauss, 1991). For there to be a gift at the end of the exchange, some unexpected extra must have been added from which the obligation to give back arises. This added dimension is linked to the identity of the donor and the nature of the relationship between protagonists. It is because this crystal was found, exchanged, or given by someone that, consequently, a part of that person is deposited in it, and so one becomes obliged. In other words, the question about what makes us give and give back or why it is that the same object may be integrated into, or leave, the sphere of trade "cannot be answered without a reference to the thing given" (Bazin, 1997: 19).

To understand the points made earlier, it is necessary to consider that a rock crystal can be treated in at least two ways. A rock crystal may be considered as a possible substitute for a series of other objects; it is a mineral that represents the class of all its fellow minerals. In this first sense, it can then be exchanged for property of equivalent value or bought at a price. In a second sense, however, when considered by itself as a single fragment, a rock crystal does not admit any substitution. It is a singular thing for which no equivalent exists. Here, we move from the register of reciprocity to that of sovereignty. Some singular crystals, such as Laurent, make a name for themselves through their ability to excite passions and arouse emotion. Within the week that followed Laurent's discovery, the news of this exceptional specimen had already travelled around the world, making a reputation for itself among the great collectors who were likely to acquire it. If economic value is in conflict with other values, it is because that particular value is itself a value like all others and not the expression of the utility of things, as economists thought for one and a half centuries. Viviana Zelizer (1994) has clearly observed that there is competition between values, as well as the presence of resistances to the extension of the sphere of the market.

What appears in this example is that the owner of the given or exchanged resource is always present in the object. However, if the donor is still present in the given thing, the "nature" of the thing itself is transformed, in the sense that it can no longer be exchanged as a vulgar object that can be substituted for another. Reducing the gift to an exchange "is simply to cancel the possibility of the gift" (Derrida, 1991: 101). As a permutation of objects considered equivalent and generalized reciprocity, the law of exchange does not recognize things in their peculiarity or in that which binds a giver and a receiver. A "prestige item" distinguishes itself specifically from a commodity or from a commodity currency as far as its meaning is not independent from the partners who exchange or possess it. What Laurent shows us is that this category comes from a construction in which a mineral finds its place only though the accumulation of the charges it records. Its singular character derives not only from the fact that it is particularly precious or unique, but also from the circumstances, contexts, and personal histories associated with it.

Two kinds of actants are at work in shaping the social life of crystals: the crystals (the things themselves) – which go from hand to hand and about which we could write a history and trace their circulation – and also the crystal hunters and collectors who compete and are involved in complex pursuits like acquiring crystals, selling them, keeping them, contemplating them, or touching them. Through their movements between crystal hunters and collectors, these crystals make and undo reputations, and they sanction and guarantee acquired or lost prestige. Giving them is, of course, a way of producing an effect (obligations, debts, and so on). But keeping these crystals incommunicado in a private collection or, on the contrary, exhibiting them for the admiration of crowds in museums and exhibitions is another means, just as effective, of achieving the desired effect of affecting social life itself.

If, as we have seen, crystals can integrate into the economic sphere, they can leave it too. At any time, crystals and minerals may leave the market sphere and turn into what one might call "prestige goods" (Gallay, 2013; Godelier, 1969; 1982; 1996; Goody & Tambiah, 1973; Mauss, 1991; Panoff, 1980; Testart, 2007; Weiner, 1985; 1992). That is to say, they will lose their use value and acquire a sign value autonomous from trade transactions, even though their acquisition may have initially resulted from a trade transaction. The acquisition of a specimen by a museum is an example of such a transformation. In such a case, the stone is deliberately removed from the network of trade exchanges, thus preventing the erosion of its value through the introduction of new minerals in the economic system (by the production and discovery of new deposits or by the exchange of new objects in the economic mineral market). Some crystals come out of the economic market to integrate into social and political spheres, either in private or public (collections), while others, like Laurent, are directly integrated into these fields as soon as they are discovered.

Prestigious crystals that were removed from the market may also eventually re-enter the economic market, partially losing their sign value as they acquire or regain use value. This use value – in brief, their price – can, however, be connected to their value as a sign, to their prestige, and to their reputation. However, the crystal can also be devalued because other discoveries have been made since the mineral was removed from the market. A piece that was once regarded as the most beautiful in its category, or even unique, can later be put in competition with other, even finer, specimens. Such is the case with the beautiful, rare, red fluorites, found in 2008 in the Mont Blanc massif, which were sold at a very high price at the time (in one case, to the Museum of Mineralogy of Chamonix). In 2009 and 2010, other red fluorites of beautiful size and intense colour were discovered, suddenly devaluing the red fluorites acquired and put into collections in 2008 and spurring a debate over the price of minerals and the value of existing specimens in the field of mineralogy.

Conclusion: The Test of Prestige

The value and prestige of minerals emerge from the exchanges and symbolic productions in which they become involved, alongside their users, as carriers of social distinction and singularity. In contemporary societies, semiologists have introduced the concept of "sign" (Baudrillard, 1972) to define the real value of goods, which are less defined by their exchange value than by the social functions (classifying individuals and groups) and symbolic ones (providing signs of status and identity) that the sign fulfils in a way similar to that fulfilled by "prestige goods" (Bonte, 2006).

If selling and buying have become the dominant activities of Western societies, and if crystals do not escape this tendency, we have to admit that these objects have not totally broken the link that connects them to people. The price transfers the use value, but a part of these crystals remains inalienable from its discoverer, even sometimes from a prestigious collector who eclipsed the discoverer by buying the mineral at full price, even before other collectors and crystal hunters had heard of it. Crystals' prices, in other words, are not enough to express their value. Something in them is deposited from the might that is the source of their power and value. Finally, crystals, especially the valuable ones, appear to be a concrete manifestation of the symbolic and imaginary dimensions of social relationships and the battles of wills shaping the contexts in which they move. What must be understood here – something that is largely unknown in the analysis of the value of minerals as art objects generally – is that values are imposed on all, and their power lies precisely because they are experienced as very emotionally invested imperatives (Appadurai, 1986; Coquet, Derlon, & Jeudy-Ballini, 2005; Gell, 1998; Heinich, 1993; 2017). In this regard, the story of Laurent, the specimen purchased in 2009 by the National Museum of Natural History in Paris, is especially relevant: its discovery and subsequent immediate attempts to see it, touch it, and assess it; its reputation and the competition and conflict over its acquisition; the controversy over the price that was paid for it; the rivalry, envy, enmity, and ambivalence revealed alongside the glory that went to the specimen's discoverer; and so on. All attest to the greatness and prestige associated with the object and its owner.

Thus, some minerals can be seen as an extension of the people who found them, performing an "abduction of agency."[2] From this point of view, we may not consider them as things, but rather "as persons" having effects, "people-objects" having kept the "faculty of persons to act" (Appadurai, 1986: 4; see also Gell, 1998; Heinich, 1993). The production, the traffic, and the value of minerals depend on the social context and the system of relationships in which they occur. This emphasis on the relationship enables us to keep in perspective the boundary between crystal hunters and minerals, between human beings and non-human beings, by contextualizing it, extending the notion of agency not only to the animal world but also to the mineral world. Converging programmatic leads have been suggested over the last few years by Tim Ingold (1986), Marilyn Strathern (1988), Alfred Gell (1998), Bruno Latour (1999), Eduardo Viveros de Castro (1996; 2009), and Philippe Descola (2005). However, it is necessary to underline that this singularity and this agency of crystals must be resituated in the categories and classes to which they belong. Indeed, minerals are never thought of as isolated but are instead placed in their category and their family of belonging. Crystal hunters, collectors, and mineral lovers distinguish themselves in their knowledge of classifications and their ability to recognize minerals in the profusion of classes and categories enabling the identification of minerals taken from the field. The most prestigious crystals, those like Laurent that make a name for themselves and stand out, are precisely the ones whose particular characteristics distinguish them most from their generic category. In other words, crystals are never considered in a purely individual way, but as entities belonging to collections of minerals. The ultimate acquisition of Laurent by the National Museum of Natural History in Paris appears not only as an element of individual or collective strategy but also as a mode of institutionalization, effectively withdrawing a mineral from the market and protecting it as a "cultural property of major patrimonial interest" (see the law of 1 August 2003 concerning sponsorship), thereby allowing it to enter the National Museum's collection and making it inalienable. The value and prestige of crystals thus depend on several factors and are not based only on their objective material qualities. From this point of view, the minerals are close to the objects analysed by Nicholas Thomas when he notices that "[o]bjects are not what they were made to be but what they have become" (Thomas, 1991: 4).

Crystals like Laurent have a dual nature. They are objects of commercial and social exchange on the one hand; on the other hand, they are people-objects, singular minerals endowed with agency, and categorized objects belonging to classes and families of minerals. We have seen that crystals can enter the commercial sphere but they can also leave it, and vice versa. The movement of these goods is meaningful precisely because some of them do not circulate and are put into a collection. So, the value of Alpine crystals is less about their exchange value than about the social and symbolic functions they serve in building ties of recognition and reciprocity. The agonistic dimension, the risk, and the value of physical commitment to the activity, reputation, honour, rivalry, and confrontation – relying on the discovery and circulation of minerals, as well as the signs attached to them – all reflect the prestige of their owners. The forms of circulation, value, and appropriation of crystals, which are reduced neither to their use nor to their exchange, suggest a resemblance to what anthropologists usually call "prestige goods." These "prestige goods" draw attention to the individuals who hold them and open the theatre of competition, representing social positions, identities, statuses, and other indicators of social and political differentiation. Rivalry and conflicts in the social spaces of crystal hunters and mineral collectors are articulated with the value of crystals, which itself plays a part in the circulation of affects and the mutuality of people and minerals.

Acknowledgments

I would like to thank the editors, Elizabeth Ferry, Annabel Vallard, and Andrew Walsh, for the long and friendly work of coordinating and adjusting the various texts of this book, as well as for the organization the Wenner-Gren Workshop on the Anthropology of Precious Minerals in Toronto, which led to this publication.

NOTES

1 "Oven" is a native term for a natural cavity from which quartz crystals, and possibly other minerals, are extracted.
2 The agency is "relational and always fits into a context" (Gell, 1998: 27).

REFERENCES

Absi, P. (2003). *Les ministres du diable: Le travail et ses représentations dans les mines de Potosí, Bolivie*. Paris: L'Harmattan.

Appadurai, A. (Ed.) (1986). *The social life of things: Commodities in cultural perspective*. Cambridge: Cambridge University Press.

Baudrillard, J. (1972). *Critique de l'économie politique du signe*. Paris: Gallimard.

Bazin, J. (1997). La chose donnée. *Critique, 596–97*, 7–24.

Bazin, J., & Bensa, A. (1994). Les objets et les choses: Des objets à "la chose." *Genèse, 17*, 4–7. https://www.persee.fr/doc/genes_1155-3219_1994_num_17_1_1257

Bonnot, T. (2002). *La vie des objets: D'ustensiles banals à objets de collection*. Paris: Editions de la MSH.

Bonte, P. (2006). La notion de "biens de prestige" au Sahara occidental. *Journal des Africanistes, 76*(1), 25–42. https://journals.openedition.org/africanistes/176

Canac, R. (1980). *L'or des cristalliers*. Paris: Denoël.

Coquet, M., Derlon, B., & Jeudy-Ballini, M. (Eds.). (2005). *Les cultures à l'œuvre: Rencontres en art*. Paris: Biro éditeur.

Derrida, J. (1991). *Donner le temps: 1. La fausse monnaie*. Paris: Galilée.

Descola, P. (2005). *Par-delà nature et culture*. Paris: Gallimard.

Gallay A. (2013). Biens de prestige et richesse en Afrique de l'Ouest: Un essai de definition. In C. Baroin & C. Michel (Eds.), *Richesse et sociétés, collection "colloques" de la maison archéologie & ethnologie* (pp. 25–36). Paris: De Boccard.

Geffray, C. (2001). *Trésors: Anthropologie analytique de la valeur*. Strasbourg: Arcanes.

Gell, A. (1998). *Art and agency. An anthropological theory*. Oxford: Clarendon Press.

Godelier, M. (1969). La "monnaie de sel" des Baruya de Nouvelle-Guinée. *L'Homme: Revue française d'anthropologie, 9*(2), 5–37. https://doi.org/10 .3406/hom.1969.367046

Godelier, M. (1982). *La production des grands homes*. Paris: Grasset.

Godelier, M. (1996). *L'énigme du don*. Paris: Grasset.

Goody, J., & Tambiah, S.J. (1973). *Bridewealth and dowry*. Cambridge: Cambridge University Press.

Gregory, C.A. (1982). *Gifts and commodities*. London: Academic Press.

Heinich, N. (1993). Les objets-personnes. Fétiches, reliques et œuvres d'art. *Sociologie de l'art, 6*, 25–55.

Heinich, N. (2017). *Des valeurs. Une approche sociologique*. Paris: Gallimard.

Ingold, T. (1986). *The appropriation of nature: Essays on human ecology and social relations*. Manchester: Manchester University Press.

Kopytoff, I. (1986). The cultural biography of things: Commoditization as process. In A. Appadurai (Ed.), *The social life of things: Commodities in cultural perspective* (pp. 64–94). Cambridge: Cambridge University Press.

Latour, B. (1994). Une sociologie sans objets? Remarques sur l'interobjectivité. *Sociologie du Travail, 34*(4), 587–607. https://doi.org/10.3406/sotra.1994.2196

Latour, B. (1999). *Politiques de la nature. Comment faire entrer les sciences en démocratie*. Paris: La Découverte.

Leach, J., & Leach, E. (Eds.). (1983). *The Kula: New perspectives on Massim exchange*. London: Cambridge University Press.

Lévi-Strauss, C. (1991). Introduction à l'œuvre de Marcel Mauss. In M. Mauss (Ed.), *Sociologie et anthropologie* (pp. 1–52). Paris: PUF. (Original work published 1950)

Malinowski, B. (1989). *Les Argonautes du Pacifique occidental*. A. & S. Devyver (Trans.). Paris: Gallimard. (Original work published 1922)

Manandhar, S. (1998). Bijoux et parures traditionnels des Newar du Népal: Une approche anthropologique et historique. Doctoral dissertation, Université de Paris X – Nanterre, 2.

Mauss, M. (1991). Essai sur le don. In M. Mauss (Ed.), *Sociologie et anthropologie* (pp. 145–279). Paris: PUF. (Original work published 1950)

Meyer, F. (1988). *Gso-ba rig-pa: Le système médical tibétain*. Paris: Editions du CNRS.

Panoff, M. (1980). Objets précieux et moyens de paiement chez les Maenge de Nouvelle-Bretagne. *L'Homme, 20*(2), 5–37. https://doi.org/10.3406/hom .1980.368070

Raveneau, G. (2006). Quand l'argent sous-tend l'échange agonistique et s'attache à la mort. In G. Lazuech & P. Moulevrier (Eds.), *Contributions à une sociologie des conduites économiques* (pp. 81–92). Paris: L'Harmattan.

Simmel, G. (1987). *Philosophie de l'argent*. S. Cornille & P. Ivernel (Trans.). Paris: PUF. (Original work published 1900)

Strathern, M. (1988). *The gender of the gift: Problems with women and problems with society in Melanesia*, Berkeley: University of California Press.

Testart, A. (2007). *Critique du don: Essai sur la circulation non marchande*. Paris: Errance et Syllepse (Matériologiques).

Thomas, N. (1991), *Entangled objects*. Cambridge, MA: Harvard University Press.

Viveiros de Castro, E. (1996). Os pronomes cosmólogicos e o perspectivismo ameríndio. *Mana*, 2(2), 115–44. https://doi.org/10.1590/s0104-93131996000200005

Viveiros de Castro, E. (2009). *Métaphysiques cannibales*. Paris: PUF.

Walsh, A. (2003). "Hot money" and daring consumption in a northern Malagasy sapphire-mining town. *American Ethnologist*, 30(2), 290–305. https://doi.org/10.1525/ae.2003.30.2.290

Weiner, A. (1985). Inalienable wealth. *American Ethnologist*, 12(2), 210–27. https://doi.org/10.1525/ae.1985.12.2.02a00020

Weiner, A. (1992). *Inalienable possessions: The paradox of keeping-while-giving*. Berkeley: University of California Press.

Zelizer, V. (1994). *The social meaning of money: Pin money, paychecks, poor relief and other currencies*. New York: Basic Books.

PART TWO

Mineral Connections

Making Preciousness: Distinction and Refraction

E L I Z A B E T H F E R R Y

The chapters in part two focus on meaning-making with precious minerals once they have moved some distance away from their source, in interaction with people and things not directly tied to their extraction. The source of minerals (or at least the fact that they have a source in the earth beyond and below the "human realm") never goes away entirely, however, which accounts for the lingering value of precious minerals, even when chemically identical equivalents can be produced in a laboratory (Walsh, 2010). But the movements of minerals into new places are more immediately generative of particular forms of value – rarity, brilliance, refinement, preciousness – than in the previous section.

What Makes Precious Minerals Precious?

What gives particular stones – diamonds, emeralds, jade, corundum, gold, and a very few others – the capacity to convene such complex practices as the making of glamour, allure, and attraction described in the following three chapters? In these brief remarks, focused on the circulation and valuing of precious minerals, I suggest that one story we can convincingly tell in answer to this question is that those stones called precious, through their material and historical qualities, enter into three dimensions of human meaning-making: the natural, the social, and the transcendental. Precious minerals engage these dimensions in particularly fruitful and complex ways.

The Natural

The category of the natural and its shifting contextual and sociohistorical boundaries is enormously complex; however, in many of its denotations, the natural pertains to those material worlds seen (at least within many European traditions) as outside of the social and even the human.[1] Bruno Latour (1992) designates the simultaneous "purification" and "translation" of nature and society as the defining characteristic of what he calls "the Modern Constitution." Furthermore, the relationship between nature and society tends to attract extraordinary amounts of cultural work and contention. That distinction – and its disruption – is a fundamental aspect of valuation processes in economic, aesthetic, religious, affective, and many other realms. Minerals, as substances mined from the earth and subject to forces on vastly different scales of time and space than those to which humans are subject, engage questions of nature especially strongly and are often taken as examples of nature in its purest or most pristine form. These "modern" forms of engagement may depend on the creation of nature as pristine, a creation that can take quite a lot of work (Ferry, 2013), or on other versions or iterations of the natural, as we shall see further on.

Filipe Calvão's chapter, concerning the ways in which source and origin play into valuations of diamonds in Angola and Switzerland, demonstrates the force of the "natural" in what makes precious minerals precious. Calvão writes about the interplay between dirt and cleanliness and between occlusion and transparency in the movement of diamonds from rough to cut. Whereas, in Angolan trading rooms, "diggers argue from the standpoint of their authority vis-à-vis their immediate proximity to the earth-tainted surfaces of nature's diamonds" and display their dirty fingernails as indices of that authority, diamonds in Geneva are purified of all specific reference to their source. In Calvão's work, we see that diamonds' non-human source (in contrast to origin in a laboratory), even though the specific sources for them are frequently occluded, continues to be important in diamond valuation and, arguably, that a generalized source of "nature" is made even more important through separation from the impure sociality of war and corruption (see discussion later in this introduction). The tension between the impetus to create a transparent, ethical chain of custody for diamonds and the mystique adhering to untraceability helps to create nature as an open signifier. As Calvão notes, this tension descends to the molecular level; while some diamonds demonstrate physical characteristics that can link them to specific places, "the industry works with the consensus that the origin of diamonds is ultimately untraceable given the very natural properties of the stone."

Annabel Vallard's chapter denotes three cases of preciousness in minerals from Thailand: the Emerald Buddha; the cutting, circulation, and valuation of jewels in the royal treasury; and the interactions between archaeologists, beads, and bead collectors in Chumphon province.

Each of these instances foregrounds one of the dimensions I explore here. For example, in her discussion of those who work with corundum crystals and other coloured stones (such as emeralds), Vallard shows how these artisans distinguish the standardization of diamond valuation and their own work, which is located in an embodied attention to the stones themselves. In the case of gem cutting, the human work is seen as bringing out the stone's already present natural qualities based, in part, on the natural qualities of perception and sensation of the corundum professionals themselves. Vallard notes not only that corundum crystals, like emeralds, are valued for their origins in specific localities and the effects of these on the stones themselves (a kind of mineral "terroir" [Paxson, 2012]) but also emphasizes "the vivid passion and attachments with which practitioners engage" the stones." Here, proper valuation depends on the communication between the natural stones and the natural bodies of those who touch, look at, and shape them.

Les Field's chapter approaches the realm of the "natural" from a somewhat different angle. Vallard and Calvão, along with other authors in this volume, take preciousness as an internally consistent though undeniably messy constellation of meanings, practices, and properties. They do not explicitly seek to debunk, puncture, or undercut the worlds created through minerals (though their analyses certainly contain implicit critiques). Field's chapter, on the other hand, has a problem with preciousness, especially as related to the taken-for-grantedness of gold's natural properties and their consequences. Field begins his discussion by focusing on a genre of writing about the "histories of gold," often appearing in lavishly illustrated coffee table books or in popular histories. These narratives link the supposedly unique material qualities of gold and the supposedly unstoppable trajectory of gold's history as currency and material for the expression of luxury, ceremony, and romantic love. Field aims to destabilize this "naturalizing" story with material from pre-European Colombian sources to show that gold's emblematic role as the pinnacle of preciousness was not inevitable and that gold's apparent agency in human history is the output of a political process: specifically, that of taking the occidental history of gold to be the only possible history, rooted in geology and human cognition and emotion. Indeed, I have seen this process in my own research on gold and finance, which has taken me into many offices of gold funds, commodities research firms, and bullion banks sheathed in luscious, glowing photographs of gold bars, molten gold, and gold leaf. These images do the work of linking gold's lustre, colour, and mass with its intrinsic, unarguable value and, thus, its aptness as a financial asset.

The Social

In his famous essay "The Spirit of the Gift" in *Stone Age Economics*, Marshall Sahlins (1972) proposes that Mauss's fundamental conception of the gift, and the hau as its motivating principle, allows for the institution of a form of sociality opposed to war or "warre," as Thomas Hobbes (2006) put it. The gift, Sahlins argues, performs for Mauss the role performed by the state for Hobbes. Importantly, in order to make this argument, Sahlins is obliged to clarify that "the war of every man against every man" (1972: 171) is itself a form of social order, albeit a destructive and chaotic one.

Participants in the contemporary evaluation and circulation of precious minerals often seek both to avoid contributing to this destructive and chaotic form of society (frequently glossed in the non-governmental organization [NGO] literature as "conflict") and to provide alternatives to it through social networks based on kinship and religion, the formation of transtemporal collectivities such as nations or ancestor-descendent ties, and through certification schemes and other apparatuses of transparency.

In keeping with his Marxist perspective, the engagement of the social and the approach to the material as congealed social relations is foregrounded most strongly in Field's chapter. The core of his objection to gold's preciousness (or at least monolithic Euro-American accounts of that preciousness) lies in the fact that it conceals the possibilities of other valuations of gold and ratifies instead a sense of gold as *inevitably* precious, without reference to human labour, history, or power (all dimensions of the social in Field's framework).

Indeed, reassertions of these dimensions of gold's preciousness form the basis for a second aim of Field's chapter. Making his essay into another kind of intervention into preciousness as a distinctive mode of agency, Field puts forward a critical enquiry into recent works in what he calls (following Bessire & Bond, 2014) "ontological anthropology," drawing a connection between a lack of recognition of the fetishism of gold in the "histories of gold" genre, described earlier, and a lack of recognition of the fetishism of objects in general in some scholars working in this literature. This provocative argument uses a concrete case (that of the naturalizing ideology of gold's preciousness) to push back against the critique of the "demystification" of fetishism that Jane Bennett (2009: 248) and others have proposed. Field argues that the privileging of "vibrant matter" need not entirely displace the Marxist critique of fetishism and puts forward the "histories of gold" as an extreme case that shows the risks of doing so.

In making this argument, Field not only intervenes into debates over gold, the agency of objects, and the vitality of matter but also links the world of scholarship and that of the histories of gold in popular publications. In each of these areas, ontological distinction is privileged over fetishism; Field argues against this privileging, noting that gold's claim to special ontological status (as intrinsic value) is at the same time an emblematic instance of fetishism.

In Vallard's chapter, the capacity of minerals to create the social emerges in a different direction, through the tensions between archaeologists and local bead collectors created by a range of different semi-precious minerals. Whereas archaeologists are concerned with reconstructing the sites of past settlements and maintaining the beads in the name of the Thai nation as their appropriate steward, the collectors seek to establish personal relationships with the beads' prior owners. Vallard notes that collectors feel a "link to humans of the past, which ... is very physical, sensorial, and emotional."

Calvão's chapter addresses one of the most important dimensions of precious mineral commodity markets, that of ethical sourcing and "transparency." As stated earlier, he deftly shows some of the paradoxical elements of this field, such as when transparency exists alongside, and in some cases produces, secrecy and occlusion concerning the specific sources of diamonds. It is perhaps useful to remember that the term "transparency" came to prominence in the 1990s with the organization Transparency International, an international NGO founded in Germany in 1993, which describes itself as "the global coalition against corruption."[2] The term later came to be used in the extractive sector by groups such as the Extractive Industries Transparency Initiative and the Kimberly Process Certification Scheme. These instruments seek to replace bad sociality with good by providing tools (in some ways cognate with the processes of the gift as described by Mauss) for ethical sourcing and a reduction of conflict and corruption. Diamonds ratified as "conflict-free" stand for a "higher form" of society, above a Hobbesian state of eternal war. As part of this process, diamonds acquire a clarity, purity, and brilliance that indexically reproduce these same qualities expressed iconically in the stones themselves.[3]

The Transcendental

I chose the term "transcendental," rather than terms like "divine" or "supernatural," in order to be able to include semiotic activity that points beyond the planes of either the natural or the social into realms that would not, however, pertain to what might be called "religion," such as aesthetic, moral, or affective realms. Some of the common material properties of minerals (lustre, crystallization, malleability, endurance), as well as their underground, underwater, and sometimes outer space origins, make them especially good conductors to the transcendental in these various forms.

We see this conductivity with particular clarity in Vallard's discussion of the Emerald Buddha, associated both with the god Indra and the Thai royal family. Vallard indicates the ways in which the Buddha, in its substance, embodies divinity. For example, she notes: "As a substitute for the same essence and the same origin as the Manijoti jewel or the 'jewel par excellence,' the statue has been endowed with supernatural abilities." The semiotic grounds for this statue merge immanence and transcendence in a particular way, one that we also find in the case of other precious minerals; as Vallard says, "the stone, then, was not considered to be the residence of the spirit but the spirit itself, the spirit and the substance being totally inseparable." We see a similar process in many contexts in which gold is used, both in religious icons and shamanic figures, in places like medieval Europe and pre-Columbian South America and also in monetary theory, in which gold is taken not as the sign of value but as value itself (Maurer, 2005). Elsewhere, I have described this phenomenon as the semiotic claim that gold (and, in this case, the Emerald Buddha) is not a sign, based on what Vallard describes as the inseparability between spirit and substance (Ferry, 2016). This semiotic claim is one mode in which precious minerals engage the realm of the transcendent.

Gestures towards the transcendent also appear in Calvão's chapter five, where diamonds' iconicity is referenced through the qualities of purity and brilliance they invoke. This iconicity occurs especially in the trading rooms of Switzerland. The quotation from the Christie's catalogue with which Calvão begins the section of his chapter focused on these sites characterizes this sense of transcendence:

> Under normal circumstances, the value of a diamond can be defined by its carat weight, cut, colour, and clarity grade. But on rare occasions, a diamond will show something beyond, something special, an extraordinary charm that cannot be explained by words. This is when you know you have a legendary diamond in your hands. (Christie's, 2014: 256)

This legendary status "beyond" the diamond's "4 Cs" grade is, at the same time, iconically demonstrated by the qualities of clarity and brilliance (made possible by the diamond's cut). The fact that the Lesedi La Rona rough diamond did not find a buyer suggests that diamonds' capacity to act as icons of transcendence is undermined when the transformation from rough to cut is made too evident.

Field also references the transcendent. He bases his argument against the unique identity of gold as precious on the historical incidence of other substances made precious in the Americas, such as jade, pearls, and feathers. To do so, he draws on the work of Nicholas Saunders (1999), arguing that Aboriginal Americans incorporated gold and gold alloys into an "existing system of sacred brilliance" and showing ways in which contemporary Indigenous people in Colombia locate gold in symbolic and ritual systems in dynamic conversation with Euro-American modes of valuation of gold-as-money. As part of this system, Kogi and other Native peoples continue to use gold objects in shamanic and other ritual contexts, which connect humans to the realms of animals and the dead, and to interact with some gold objects as living beings (Ferry & Ferry, 2017).

In this introduction, I have proposed that one way to explain, or at least to explore, how the preciousness of precious minerals works in their circulation and evaluation is through a triangle of the natural, the social, and the transcendent. Each of these chapters demonstrates a different configuration of these three dimensions, motivated by particular historical circumstances (and their ideological affordances) and by the materialities of each mineral substance. We have placed Field's essay as the last substantive chapter of the volume to foreground this very point: that preciousness comes into being through the matter of minerals *and* through the social actions in which they engage and are engaged. Indeed, as we elaborate in the afterword, we take preciousness to be a quality of certain kinds of objects in which incommensurability and a dense and plural sociality are both particularly distilled and particularly hard to see. Precious objects are both set apart by their distinctiveness and situated at the centre of multiple, refracted modes of valuation.

The careful ethnographies in part two map the patterns and variations of preciousness in particular cases. Field's chapter adds to these mappings a bracing, critical view on what is not included and not recognized when preciousness is successfully established. Taken together, these approaches allow us to take preciousness seriously while also recognizing its ideological context, a context within which we ourselves are constantly caught up.

NOTES

1 Human beings are also recognized as being themselves creations of the natural world and, at least at certain moments, subject to its rules.
2 See the "Who We Are" section of the Transparency International website at https://www.transparency.org/whoweare/history. The slogan "the global coalition against corruption" is part of the company logo.
3 For insightful discussions of these processes in the Colombian emerald economy, see Brazeal (2016) and Caraballo Acuña (2018).

REFERENCES

Bennett, J. (2009). *Vibrant matter: A political ecology of things*. Durham, NC: Duke University Press.
Bessire, L., & Bond, D. (2014). Ontological anthropology and the deferral of critique. *American Ethnologist, 41*(3), 440–56. https://doi.org/10.1111/amet.12083
Brazeal, B. (2016). Nostalgia for war and the paradox of peace in the Colombian emerald trade. *The Extractive Industries and Society, 3*(2): 340–9. https://doi.org/10.1016/j.exis.2015.04.006
Caraballo Acuña, V. (2018). Comerciar sin afiebrarse. Experiencias sensoriales y oposiciones cualitativas en la formalización de la economía esmeraldera en Colombia. *Revista Colombiana de Antropología, 54*(2): 9–33. https://doi.org/10.22380/2539472x.459
Christie's. (2014, 14 May). *Magnificent jewels*. Auction catalogue. Geneva: Manson & Woods Ltd.
Ferry, E. (2013). *Minerals, collecting, and value across the U.S.-Mexican border*. Bloomington: Indiana University Press.
Ferry, E. (2016). On not being a sign: Gold's semiotic claims. *Signs and Society, 4*(1): 57–79. https://doi.org/10.1086/685055
Ferry, E., & Ferry, S. (2017). *La batea*. Bogotá: Editorial Ícono.
Hobbes, T. (2006). *Leviathan: Or the matter, forme and power of a commonwealth ecclesiasticall and civil*. (Vols. 1–2). G.A.J. Rogers and K. Schuhmann (Eds.). London: A&C Black. (Original work published 1651)
Latour, B. (1992). *We have never been modern*. Cambridge, MA: Harvard University Press.
Maurer, W. (2005). *Mutual life, limited: Islamic banking, alternative currencies, lateral reason*. Princeton, NJ: Princeton University Press.
Paxson, H. (2012). *The life of cheese: Crafting food and value in the United States*. Berkeley, CA: University of California Press.

Sahlins, M. (1972). The spirit of the gift. In *Stone age economics* (pp. 149–83). New York: Aldine Press.

Saunders, N. (1999). Biographies of brilliance: Pearls, transformations of matter and being, ca. AD 1492. *World Archaeology, 31*(2), 243–57. https://doi.org/10.1080/00438243.1999.9980444

Walsh, A. (2010). The commodification of fetishes: Telling the difference between natural and synthetic sapphires. *American Ethnologist, 37*(1), 98–114. https://doi.org/10.1111/j.1548-1425.2010.01244.x

4 When Stones Become Gems: Valuations of Minerals in Thailand

ANNABEL VALLARD

For a stone represents an obvious achievement, yet one arrived at without invention, skill, industry, or anything else that would make it a work in the human sense of the word, much less a work of art. The work comes later, as does art; but the far-off roots and hidden models of both lies in the obscure yet irresistible suggestions in nature.

Roger Caillois,
The Writing of Stones (1970: 2)

When the sun set in Geneva on 12 May 2015, the "Sunrise" – a 25.59 carat, faceted, cushion-shaped ruby mounted on a Cartier diamond ring – sold for US$30.3 million, over three times the previous world record price for a ruby.[1] The precious gemstone was accompanied by reports from the Swiss Gemological Institute (no. 78414) and the Gübelin Gem Laboratory (no. 15020105), both indicating its weight in carats, its Burmese origin, the absence of indications of heating, and its "pigeon blood colour." While these documents underline the gemstone's "outstanding characteristics," whose "combination" is "impressive" and "immeasurably rare" (Sotheby's, 2014), they also act as prestigious certificates of authenticity, establishing and enhancing both the origin of the ruby and its quality, here crystallized in its size, colour, and shape. Under the cover of scientific expertise, the prose in these reports was, as it usually is, "hyperbolic" and linked to marketing strategies (Walsh, 2010). The Sunrise was not just described as "superb," but as "a unique treasure of nature" (Sotheby's, 2014).

For many in the world of precious minerals and gems, stones are chiefly valued in relation to their "naturalness" (Ferry, 2005) and even their "pristineness," collectible minerals in particular being kept virtually "untouched by human hands" (Ferry, 2013: 166), sometimes in museums such as the one described in this volume's introduction. Practitioners, whether miners, gemologists, lapidaries, traders, or collectors, emphasize that "at least some of [a precious mineral or gem's] qualities [related to their aesthetics in particular] are [due to] the forces, divine or natural, that brought them into existence" (Walsh, 2010: 110). Human interventions on mineral specimens and their entanglement in previous social relations tend to be underestimated and even erased in international marketing discourses and displays (Ferry, 2005: 430–1). Nevertheless, the work of transforming rough stones into gems and jewellery is eventually recognized and esteemed by experts, laboratories, and certification agencies. The Sunrise ruby, for example, owes its "homogenous and saturated colour," both "to a combination of well-balanced trace elements in the stone, typical and characteristic for the finest rubies of Mogok," and to its "well-proportioned cutting style." The work of the cutter "further pronounced" its colour, giving it "vivid red hues due to multiple internal reflections." The ruby's natural colour and its "high clarity" were furthermore associated with its "brilliance" (Sotheby's, 2014), which is yet another result of the skilled faceting of the stone by artisans.

As this case suggests, evaluations of gemstones commonly stress a need for balance between the inherent properties of particular minerals and what humans bring to them: the preciousness of some precious minerals becomes apparent only when that balance is achieved. In this chapter, I explore this point further by following the journeys of three different sets of minerals related to the Thai kingdom and the work associated with them through "meshworks" (Ingold, 2007) that transform rough minerals into precious gems. I begin by recounting the entwined histories and ritual lives of the Emerald Buddha statue and Thai royalty. Through a discussion of the donations, devotion, and work that have gone into the production of this statue, its bejewelled garments, and other precious gems of the Royal Treasury, I deal with the potency of certain minerals, their resistance to classification, and their intertwining with the authority, shine, and fame of royalty in Thailand. While the first two sections are based largely on public accounts and previous case studies, the third section draws from ethnographic fieldwork with Thai bead collectors and French archaeologists. It reveals a different set of human-mineral potency dynamics operating in the Upper South of the country in relation with and opposition to national authorities. These three cases are developed through ongoing research dedicated to the modalities by which people operating in various sociotechnical contexts in Thailand (from mines to jeweller workshops to international brokers) experience, interact with, and organize around minerals. They highlight how the valuation of minerals as precious is a multifaceted process, not only anchored in semiotics and norms but also in bodily practices, sensual experiences, and emotions.

The Emerald Buddha

On 16 March 2014,[2] the headlines of the *News from the Palace*, a television program dedicated to the daily cultural and social schedule of the Thai royal family, offered a short report on the temple of Phra Sri Rattana Satsadaram, the residence of the Holy Jewel Buddha or Phra Phuttha Maha Mani Rattana Patimakon, the official name of the Emerald Buddha. Located in the centre of Rattanakosin Island, or the island of "Indra's precious jewel," this site was chosen as the capital of Siam (now Thailand) in 1782 by His Majesty (HM) King Rama I (1782–1809), the founder of the current ruling dynasty, the Chakri. It remains the historical, political, and religious centre of Bangkok, not least because of its association with the Emerald Buddha statue.[3] The televised news depicted His Royal Highness (HRH) Crown Prince Maha Vajiralong-korn arriving at the temple at 17:17 on the first day of the hot season to proceed with the seasonal costume change of the statue in the name of his ailing father, King Bhumibol Adulyadej (Rama IX).[4] In keeping with a standard narrative,[5] the journalist recounts how, according to a strict protocol, the crown prince replaced the winter garments of the Emerald Buddha to mark the beginning of summer on the first waning moon of the fourth month of the lunar calendar.[6] On the television screen, the crown prince, assisted by a ritual officiant, is shown performing the ceremony on a golden staircase installed behind the Buddha image, which is seated in a meditative pose upon a golden throne enshrined on a monumental five-tiered golden altar about seven metres high. This high position in a saturated environment of gold enhances both the Emerald Buddha's rather small size – only 66 centimetres high and 48.3 centimetres wide – and its strong green colour.

Three times a year – at the beginning of the cool, the hot, and the rainy seasons – HRH Crown Prince Maha Vajiralongkorn repeats the same ritual. He begins by greeting the Buddha image with a *wai*, a formal bow performed with palms joined, and offering Thai floral garlands. On the day described earlier, the television report shows him removing the winter season crown, but not taking off the delicate and spectacular winter drapery that covers the statue's neck and torso – a meshwork of interlaced enamels, gold, and garnets joined to form ancient Thai flower patterns, replicating a knitted monastic winter shawl. The statue undressed, the crown prince cleans the Emerald Buddha to the sound of ritual conches. At his side, the officiant hands him the necessary ritual objects, one after another: a conch shell–shaped pot containing floral scented water, which he gently pours over the statue, and white cotton cloths with which he wipes the holy image. The statue cleaned, the crown prince places the summer season crown, a tall, slender gold tiara sparkling with thousands of diamonds, rubies, and sapphires, on the Emerald Buddha's head. The television program doesn't show how the statue was eventually fully dressed in other elements of its summer paraphernalia, composed of body ornaments (necklaces, breastplate, belt, ring, bracelets, anklets, shoulder pads, knee adornments). Instead, it jumps directly to images of the crown prince sponging the fabrics that were used to wash the statue and collecting the holy water, which he pours gently on his own head. After worshipping other Buddha images present in the Ordination Hall in silent prayer and in a solemn mood, the crown prince sprays holy water, first upon the officials, among whom usually stands the prime minister and other high dignitaries, and then upon the few of his subjects allowed to assist in the ceremony outside the hall.

Throughout the year, the king, the crown prince, and other members of the royal family of Thailand worship many Buddha images. The Emerald Buddha, however, occupies a place like no other in the history of Thailand. Indeed, it had been chosen by King Rama I as the palladium of the kingdom and of the Chakri dynasty (Brown, 1998; Ladwig, 2000; Lingat, 1935; Narula, 1994; Notton, 1928; 1933; Peleggi, 2009; Reynolds, 1978; Rod-Ari, 2010; Roeder, 1999; Swearer, 2004; Woodward, 1997), a concrete entity ensuring their safeguard, protection, and prosperity. This choice was rooted in the long history of the Emerald Buddha as palladium of various Tai[7] kingdoms, for example, the Lao kingdom of Lan Xang (Reynolds, 1978), which anchored the statue in a prestigious lineage of royal sponsors, devotees, and hosting cities (Chiu, 2012: 60, 74). The Emerald Buddha also has a long-time association with a royal cult as guardian of prosperity and fertility for the kingdom, notably through its relation to Indra, its principal divine patron. Indra, whose green skin colour is associated in Tai arts with his generative abilities as the god of rain and thunder, assisted Nagasena, a devoted Indian Buddhist who lived about 500 years after the Buddha's final nirvana, to create this statue from a radiant jewel named Amarakata. This latter piece was given as a substitute for the Manijoti jewel of the ideal Buddhist universal king or *cakkavatti*[8] (Chiu, 2012: 63–4), who rules the world ethically and benevolently, providing a model for Tai rulers even today. As a substitute for the same essence and the same origin as the Manijoti jewel or the "jewel par excellence," the statue has been endowed with supernatural abilities, which were reinforced during its consecration when seven relics of the Lord Buddha flew into the image and animated it.

The statue acts as an agent, anchoring and "legitimating kings and kingdoms through its presence," since it is known to "come [only] into the possession of the more politically righteous and religiously meritorious king" (Rod-Ari, 2010: 47–52). The agency of the Emerald Buddha is not only related to the Buddhist tradition, which credits Buddha images with "radiance" and a "fiery energy"(Tambiah, 1982: 6) "that endows the image with a cosmic presence" (Reynolds, 2005: 216), but also to ancient cults of stone that were important long before the spread of Buddhism in Southeast Asia (Mus, 1933). According to Lingat (1935), the Emerald Buddha was carved from a singular piece of mineral, probably a stone considered miraculous that was worshipped in ancient times, and reshaped to take on the appearance of a Buddha image during the Siamese Theravada reform movement (late fourteen and early fifteen centuries).[9] In pre-Buddhist and pre-Tai times, stones were considered miraculous by virtue of the materials they were made of. Able "to grow," these stones were associated with vegetal growth, and were considered far from lifeless (Mus, 1933: 376). In Tai societies, these virtues were also attributed to the presence of some spirits, generally related to the safeguarding of local territories. The stone, then, was not considered to be the residence of the spirit but the spirit itself, the spirit and the substance being totally inseparable. Through worshipping these specific stones, inhabitants and local chiefs attempted to reconcile their "power" or "ability to affect" the world (Reynolds, 2005: 211). The more that the stones, and later the statues, were considered agents able to act among humans, the more chiefdoms entered competitions and conflicts to capture them, their aura, and their potency as a way to bring sovereignty, luck, and prosperity to themselves and their kingdom.

The Chakri kings, each in their own manner, seized the holy figure and participated in its fashioning and maintenance as the singular protector and patron of their dynasty and of the kingdom of Siam from the eighteenth century onward. Rama I commissioned the Ordination Hall that enshrines the image (Rod-Ari, 2010: 100–10). Its architecture and mural paintings were specifically designed to establish and enhance the centrality of the Emerald Buddha inside the Grand Palace, where it presides, even today, over all important public ceremonies and the king's personal devotions. More broadly still, it presides over the capital, Bangkok, and over the kingdom as a whole. The presence of the Emerald Buddha in the sphere of influence of the Siamese kings was not merely physical, manifested in its enshrinement. It was also conceptually entangled in the history of their realm and dynasty. The fourth Chakri king, King Mongkut (1851–1868), rewrote the official history of the gem, taking inspiration from the Western science emerging in Siam during that period. He officially criticized the legendary origin of the Emerald Buddha and claimed that the statue was made from Chinese jade[10] (and not from a mythical gem given by gods) and carved in the fourteenth century following the aesthetic of the Tai Chieng Saen school (described by Lingat [1935: 13] as an ancient Lao school).[11] By doing so, King Rama IV recentred the holy image within the history of northern Tai kingdoms and his kingdom itself within a larger Buddhist "sacred geography" (Rod-Ari, 2010: 52).

After King Rama I decided to place himself and his kingdom under the Emerald Buddha's protection and patronage (Lingat, 1935: 23), his heirs went further, proceeding to an exclusive sensorial rapprochement with the statue. Stylistic studies suggest that the Emerald Buddha was never intended to be seen without ornaments, that is, "nude" (Rod-Ari, 2010: 65). When the statue was in Vientiane, the capital of Lan Xang before its capture by the Siamese, its ornaments were changed twice a year (Reynolds, 1978: 184). At the arrival of the statue in Bangkok, the Chakri reinterpreted these ancient rituals, common in Lao cults but not so much in Bangkok (Lingat, 1935: 25). The changing of the Emerald Buddha's seasonal costume by the king himself (assisted by palace ritual officiants) became central in the process, as did the cleaning, an act that evoked the Lao auspicious practice of sprinkling lustral water onto holy statues. To celebrate the arrival of the statue, King Rama I commissioned new costumes, one for the hot season and another for the rainy season. A third costume for the cold season was later ordered by King Rama III (1824–1851), "giving him an additional opportunity to have physical contact with the icon" (Rod-Ari, 2010: 174). Apart from this ceremony and the king himself, no one is authorized to approach the holy statue, let alone touch it. This practice contrasts deeply with other forms of worship involving statues, specifically in northern Tai ritual spaces where devotees might, on certain occasions, pour lustral water on statues themselves.[12]

This exclusive relation with the palladium, fashioned through a ritualized and sensible rapprochement between the Chakri kings and the holy gem, corresponds with a separation of these two from the larger community of devotees. While the Emerald Buddha, as royal palladium, was an icon exclusively in use at the court (Rod-Ari, 2010: 66), the statue was occasionally seen in public. But, for security reasons, King Mongkut forbade its transportation outside the palace and, in times of crisis, in ritual processions around Bangkok.[13] This prohibition continues today. Even for the Rattanakosin bicentennial celebrations in 1982, the statue was not moved outside the holy hall. Its "twin" image (Woodward, 1997) – the Phra Buddha Sihing – was displayed during the royal barge procession down the Chao Phraya River as a substitute for HM King Rama IX (Tambiah, 1982: 16).

While the statue and the king seem inseparable, both are subject to intense intertwined devotions. In 1996, for example, thousands of laymen participated through donations of money, gems, and precious minerals in the fabrication of a set of perfect replicas of the statue's costumes that were commissioned by the Royal Household Bureau and the Treasury Department for HM King Rama IX's jubilee celebrations, marking the fiftieth anniversary of his accession to the throne (Werly & Maquin, 2004; see also the next section). This close physical entanglement and the linked destinies of the statue and King Rama IX may eventually be read as a premise for his "potential for divinity" as a "post-mortem guardian [deity]" of the territory, a potential that those in ancient Tai kingdoms' "royal line clearly possess," but not all may achieve (Reynolds, 1969: 83).

Through historiography, architecture, arts/crafts, and rituals, the Chakri dynasty mobilized the Emerald Buddha – the "perfect gem" of Buddhism – to elevate it as the "axis-mundi"of the Theravada world (Rod-Ari, 2010: 127). As such, they created a "well-orchestrated program to establish Siam as a modern [Buddhist] center" (127) and contributed to making the realm the archetypal Buddhist kingship and establishing the figure of its king as a potential *cakkavatti*, a universal world ruler, or a *thammarat*, a moral ruler governing according to the law of the Dhamma (Askew, 2002; Roeder, 1999: 17). Intrinsically linked with the good fortune of the dynasty, the kingdom, the king himself, and the potency to govern, the statue has been the subject of controversies. During periods of political turmoil, the integrity of the Emerald Buddha is often contested. Stories of fakes, theft, and corruption of its sacred potency have been periodically evoked to explain disruptive instabilities in the kingdom. After 1932 and the revolution that led to the replacement of the absolute monarchy by a constitutional monarchy, some high dignitaries noted that the Buddha may have lost its capacity to act supernaturally (Lingat, 1935: 24) and even that the statue might be a replica made of the same materials, like the Emerald Buddha located in Lampang, which was initially intended to substitute for the original in times of danger (Lingat, 1935: 29, quoting Prince Damrong). While the presence of the holy image in the kingdom indexes the political sovereignty of the king (Tambiah, 1982: 16), its ontological status seems to always be in question. Should it lose its status as a concrete manifestation of potency, it risks being reduced to an icon (a representation by resemblance) or a mere symbol (representation by "arbitrary" social convention). To attest to and confirm both the Emerald Buddha's authenticity and the king's potential to be sovereign and progress towards divinity, the Chakri have emphasized the concrete materiality of the statue as something to which only the king himself has access through rituals. Only this intimate contact and exclusive periodic proximity can sustain the capacity of the statue and the king to affect the world through the interweaving of political and religious leadership prescribed by Buddhist kingship. This process clearly involves more than just the king and his attendants as key actors. What remains to be studied is how these scarce minerals with distinctive properties, which also play key roles in this process, afford not only iconic and symbolic representations of authority but also concrete sensorial and emotional rapprochement that may, paradoxically, help to achieve the Buddhist ideals of renouncing the material, sensorial, and emotional world.

Royal Jewels, the Standards of International Gem Markets, and the Inherent Ambiguity of Stones

From rough minerals to socially valued gems, stones require considerable semiotic as well as material work to reveal their potential and, eventually, to remove the uncertainty that weighs on them. This requirement is certainly the case with the Emerald Buddha and also with a jewel donated to HM King Rama IX by a group of wealthy Thai businessmen on the occasion of his golden jubilee in June 1996. Initially described as a large golden topaz, the "Unnamed Brown" was in fact a 546 carat yellow-brown diamond from South Africa, now known as the Golden Jubilee Diamond. While the stone was originally meant to be set in the king's royal sceptre or in the royal seal of Thailand, it is instead kept unset among the crown jewels of the royal family and on display in the Throne Hall of the Royal Museum at Pimammek Golden Temple in Bangkok. The Golden Jubilee Diamond has no market price; however, specialists estimate that it would fetch more than US$12 million at auction. Given the great value of the stone, authorities temporarily concealed its quality in order to prevent Thai citizens from becoming upset over such extravagance at a time when the country was facing a financial crisis (Barnes, 2016), but the truth was eventually revealed without any damage to the king's royal image. That this and other exceptional stones in the Royal Treasury don't rouse protests or controversy in a Buddhist kingdom undermined by social inequalities is possibly linked to the regimes of political authority prevalent in the Thai social order. In the Theravadin Buddhist cosmology, each being navigates through successive rebirths determined by merit or demerit accrued in past lives. In the former moral-political system, the *cakkavatti* king was – and still is – at the apex of this social order. His position is considered a testament to his accumulation of "virtue," "moral perfection," and "charisma" (*barami*) and an indication of his ability, in a context of patron-client relationships, to maintain a large network that will reinforce those qualities (Jory, 2002). In this cosmology, physical beauty, as well as wealth and status, are considered material evidence of the karmic inheritance. Jewellery associated with clothes in gold, silver, and silk visually serve to fashion personal images and reinforce their meritorious aura (Peleggi, 2002; Koizumi, 2009; Woodhouse, 2012; Graham, 2013). Through gifting and working these substances, donors and artisans also generate virtue and merits for themselves under the umbrella of the authority figures involved, whether they are kings, monks, or Buddha statues (Tambiah, 1978). Thus, the Golden Jubilee Diamond and other contents of the Royal Treasury speak to the values attributed to their material properties, to the desires of those who give and create them, and to the potency and authority of the royalty and statues for whom they are destined. These attributed values are especially apparent in the collaborations that recently produced new sets of bejewelled garments for the Emerald Buddha discussed in the previous section.

Nearly two hundred Thai craftsmen were required to fulfil the prestigious task of producing the Emerald Buddha's new garments. Among them were "royal goldsmiths" affiliated with the Bureau of the Royal House of Thailand, artisans whose work and positions were revived after the 1995 creation of the Golden Jubilee Royal Goldsmith College with the support of HRH Princess Sirindhorn, daughter of HM King Rama IX. In the early 1990s, the knowledge of ancient goldsmithing techniques had faded away to the point that goldsmiths familiar with the techniques of the Ayutthaya and early Rattanakosin periods, who could make – or restore – ornaments and utensils used exclusively in royal ceremonies, were extinct, according to Niphon Yodkumpun, a "royal goldsmith" himself (Dharabhak, n.d.). The task of creating a new set of costumes provided a unique opportunity to train new goldsmiths in traditional techniques, requiring the trainees to study the old paraphernalia, record their shapes on paper, and absorb the original settings of diamonds, rubies, sapphires, and other stones composing them, as well as the gold and enamel work done centuries ago (Werly & Maquin, 2004: 12). The challenge was great: replicate as closely as possible the original design with new materials, metal carving, and assembling technologies, each piece bearing a historical and religious meaning. The quantity of materials involved in the project was enormous, and the size – and commercial value – of some stones were "augmented"; cabochon stones, in particular, were far bigger than in the original designs. In its July 1996 issue, the magazine *Bangkok Gems and Jewellery* noted that 1,440 pieces of April-cut diamonds, weighing 162.20 carats, and 15,719 sapphires, rubies, and other stones, totalling 3005.94 carats, were required for the winter costume alone (quoted by Werly & Maquin, 2004: 76). Gem and jewellery traders from all around the country donated stones, gold, and other minerals required to proceed as meritorious acts for King Rama IX's jubilee.

Goldsmiths, lapidaries, and other practitioners worked each of the minerals involved individually before combining them with others to compose the costumes. All stones went through long chains of human association and cooperation along which their roughness was progressively worked, transforming opacity into transparency, dullness into shine, and rough stones into cut and polished gems. Throughout this process, common to the production of commercial diamonds, rubies, sapphires, and other gemstones, the cutter is often seen as a key agent who reveals the potential of minerals. When the "Murky Brown," the other name for the "Unnamed Brown," was first discovered, for example, nothing suggested its fabulous destiny. Although impressively large, as a rough diamond it was also strongly imperfect, its surface covered in cracks and its interior riddled with numerous inclusions (De Beers, 2008). Cutting it required two years of effort from Gabriel Tolkowsky, a sixth-generation diamond cutter and the great nephew of Marcel Tolkowsky, the creator of the modern round brilliant cut. The process included thousands of hours of careful observation without altering the stone, contemplating the rough's shape, its colour, and the inclusions present in its depths, potentially affecting both its toughness and its potential brilliance. It also required the creation of hundreds of full-sized copies made out of transparent resin, used to test the technical processes that would allow the greatest weight retention of a mineral whose fracturing would be decisive (Tolkowsky, 2001). The cutter had to navigate through many physical obstacles to work the stone into a mature, 148 facet "fire-rose cushion cut" diamond – where a rose cut usually releases only up to 24 facets – revealing its "warm" colour and its "mysterious shine" (De Beers, 2008). Unexpectedly, the cut intensified its colour, and the "Ugly Duckling" emerged as a "Magnificent Swan," whose transformation was completely fulfilled when presented to HM King Rama IX after having been blessed by Pope John Paul II, the Buddhist Supreme Patriarch, and the Islamic Chularatchamontri of Thailand (Tolkowsky, 2001).

Lapidaries are particularly careful to ensure balanced proportions and the symmetry of emerging gems, wary of under or overcutting when playing on their surfaces, angles, and thickness. The passage of light through hard stones and its refraction in their inner volume fundamentally affect human visual perception and directly influence the gems' valuation. To achieve this balance and symmetry, practitioners rely on the expertise and sensibility of their hands, the touch of their skin, and the sharpness of their eyes, trained through long years of practice. Various gauging and scaling instruments assist them to focus their movements and to homogenize the faceting. While cutting work is not an individual task – except for some exceptional stones – cutters are continuously adjusting and altering their approach to minerals, the properties of the material at hand never being totally and definitively seized at once. Open to the eventual emergence of material variations, invisible or unperceived at first, cutters demonstrate a strong sensibility for the affordances of the matter (Gibson, 1979; Norman, 1999; Knappett, 2004; Naji, 2009a; 2009b) and to its sometimes continuously changing "possibilities for action" (Ingold, 2000). In this close intimacy with minerals, they are also "dynamically shape[d]" as subjects, as efficacious actions on matter are also efficacious on the human body, self, and socialization (Naji & Douny, 2009: 418).

In the end, each mineral of the costume, facetted and polished according to its own singular properties, has been transformed and come to be labelled as *cut*. Along their trajectory, the gems have been repeatedly considered by professionals engaged in mining, cutting, polishing, and marketing through various valuation charts, evaluated according to their size, colour, clarity, shape, or origin. But, unlike "colourless" diamonds, "fancy coloured diamonds" and other coloured stones, corundum in particular (sapphire and rubies, which are endemic in Southeast Asia), are not graded within a common and standardized system such as the 4Cs (see the introduction to this volume) or national price charts, which are supposed to limit but not to erase (Calvão, 2015; see also chapter five of this volume) discussions over their identification and pricing. At each step, the trade in corundum minerals involves intense negotiations over their quality through which traders attempt to stabilize their commercial value, at least temporarily. Of course, the valuation of some precious minerals is complicated by internal inclusions that affect particular stones' density, texture, and colour (Brazeal, 2017). Distinct from the substance in which they are embedded, such inclusions are known to be good signatures of stones' geographical origin to expert evaluators. They also contribute to the variability of a stone's properties, however, leading to almost endless possibilities for qualification and valuation.

In this uncertain world, some specialized gemologists, such as Richard Hughes (1997), have attempted to develop guidelines and vocabulary to standardize the description, grading, and pricing of coloured stones (notably based on quality in terms of colour, clarity, cut, weight, and market factors). Referring to his approach to the problem, Hughes explained that he faced some objections from traders afraid of losing their advantage as exclusive connoisseurs of minerals' qualities and value. The traders he had met were especially concerned that such formalizing of expertise over minerals would depersonalize, desensorialize, and disqualify their existing engagements with stones, relations based on the "radical" perspective ("in the age of high-powered microscopes") that "to simply *look at the gem*" is enough, provided the person doing the looking has the right experience (Hughes, 1997). While colourless diamonds can be traded, they said, with only their certificates "in an indiscriminate manner, in some cases without ever viewing the gem," the trading of coloured gems requires that people with knowledge and training handle them (Hughes, 1997).

While these specialists underline the powerful "attachment" (Hennion, 2004) that coloured minerals may arouse in those who handle and/or possess them (see the next section), they also prove that, apart from the pure pleasure inclusions offer to "amateurs" (Hennion, Maisonneuve, & Gomart, 2000) who contemplate the internal richness of natural stones, practitioners have to develop their "sense of things" (*"art de la prise"*) to qualify and value the stones, which involves a series of tests between their own "landmarks" (*"repères"*) and contingency "folds" (*"plis"*) that articulate normative conventions, social networks, techniques, and also their own sensory/sensitive experience (Bessy & Chateauraynaud, 1995). What these practitioners value the most in this process seems to be the close contact with the disruptive capacity of the material to affect their senses and perception, and its capacity to "hold" them, to "surprise" them, and to attract them more than expected (Hennion & Teil, 2004). While standardized charts may help to qualify and value a particular stone on the market, only the intense sensory experience and the application of "educated attention" (Ingold, 2001) and/or "educated emotion" (Hughes, 2001; see also the introduction to this volume) to the surface and thickness of this durable material and its paradoxical constant uncertainty may ultimately maintain the vivid passion and attachments with which practitioners engage.

This close contact with the stone is precisely what the artisans who created the replicas of the three bejewelled seasonal costumes of the Emerald Buddha could not fully experience, as nobody except the king himself is allowed to touch the palladium of Thailand that they were supposed to magnify. The magnificent creation made through the assembly of thousands of independently worked gems was accomplished at a certain distance from the Emerald Buddha on a same-size replica modelled after the original (Werly & Maquin, 2004). To visually intensify the presence of the statue, these artisans created gold settings and carefully arranged diamonds and coloured stones in delicate patterns that would maximize their combined optical effects to produce expected brilliance, lustre, and fire. These qualities are sought after in Buddhism as they testify to the enlightenment of Buddha, whose physical appearance is of a "splendid golden color," his body's radiance, purity, and incorruptibility being signs of his moral and spiritual perfect achievement (Swearer, 2004: 134). While revealing the physical potentialities for radiance of the stones and their assemblage, these skilled makers also infused the gems with their dexterity as artisans, employing ancient techniques and their devotion as Buddhists engaged in meritorious acts. Nevertheless, their intimacy with the Emerald Buddha could not go further. Artisans cannot dress the statue. This privilege is restricted to the time and space of the ritual and its performance by the king. Solely on occasions like the one described earlier do gold, enamels, and other rare and shining stones aggregate with the material of the statue itself, becoming part of its body image, eventually its ontology, and enhancing its *barami*. In the royal complex, this heteroclite mineral assemblage acts as a field of cosmic vibrancy linking the statue, the king, the monastic community, the audience, the donors, the craftspeople who proceeded over the creation of the costumes, invisible entities of the territory, and, ultimately, previous manipulators of the statue whose charismatic aura might be "sedimented" in its inner matter (Tambiah, 1984). This mineral saturation, encompassing not only the Emerald Buddha statue itself and its bejewelled costumes but also the golden altar where the statue is installed and other Buddha images it presides over in the royal chapel that heighten its vitality (Bennett, 2008), acts through its sensorial effect of radiance upon humans, deities, and the cosmos as a whole, which it contributes periodically to reorder.

By capturing the donations of his royal subjects and Buddhist devotees, and through the creative work of royal goldsmiths that he commissioned, the king on this occasion reinforced the Buddhist worship community under his patronage and authority. As royal patron, he only partially delegated his attribute as outfitter of the Buddha's costumes, which are the closest things that touch the holy statue apart from him (and some high-ranking officiants) in a context where vital components of beings affect the substances of clothes and ornaments of those who wear them. He is, indeed, the only one able to concentrate the materials and the people as well as to sustain, due to his own *bar-ami*, the aggregation of the fierce mineral energy the audience may only contemplate from below and at a reasonable distance, the axis mundi of Theravada Buddhism – as well as the king himself – being unreachable for mundane contact. If, following Swearer, we consider that "the presence of the Buddha is latent or potential" in Buddha's statues and that only a royal monarch is sufficiently powerful "to actualize (his) presence in the mundane world," notably by the manipulation of his relics (Swearer, 2010: 109), the ritual of the seasonal changing of the Emerald Buddha's costumes underlines the statue's ontological continuity with the king's body and his status as *cakkavatti* king in becoming – "the Buddha in his cakkavatin aspect" (Swearer, 2010: 109) being figured not only by its untouchability but also by the stylistic characteristic of its three costumes representing the "dual power of monk and king" (Swearer, 2004: 194).[14]

When combined, these bejewelled seasonal costumes and the Emerald Buddha represent a kingdom that comprises both the devoted and the figures of authority. The donations, collection, and work of precious minerals that make these costumes possible are not simply acts of generosity, evaluation, or skilled artisanship but also acts of devotion carried out in the name of the royal patron. It is only the king, however, who is able, through close physical contact with both these garments and the Emerald Buddha, to realize the purpose of all this work, enabling it to radiate over larger scales than any individual member of the community of devotees could reach: the scale of the Tai Buddhist kingship, the Thai nation, and, eventually, the Theravada Buddhist world as a whole.

Ancient Beads, Patrimony, and Intimacy

Thai cosmic politics involve the manipulation, collection, and display of other rare and potent minerals.[15] Beyond national authorities officially in charge of fine arts and antiquities, the royal family is credited with being custodians and promoters of Thai culture, history, heritage, and, as a consequence, archaeological remains. They owe their status in this regard to the ancient practices of the antiquarians of their predecessors (Peleggi, 2007; 2013) and to the legacy of late nineteenth century histories of the kingdom of Siam, which privileged court chronicles and dynastic records, leading to enduring visions and versions of the past dominated by Thai kings and their accomplishments (Jory, 2011). HRH Princess Maha Chakri Sirindhorn, who holds two master's degrees in oriental epigraphy and arts, as well as a doctorate in science of education, singularly incarnates this royal cultural aura. She is credited with great intellectual and moral authority, not only due to her karmic birth but also to her education, and has been, as a consequence, the recipient of an outstanding donation. On 2 September 2012, her authority in this regard was highlighted in the *Bangkok Post* under the headline "Princess to Receive Ancient Beads." The day before, a couple from Chumphon province announced on television their intention to donate thousands of prehistoric artefacts unearthed from their land to the royal family, along with five *rai* – about 0.8 hectares – of land for the construction of a museum to house these treasures. In fact, the story began seven years earlier when the husband dreamt of a man from the past who urged him to dig in their fourteen *rai* land for treasure. He and his wife did so with great success. Their parcel of land proved to be exceptionally rich in beads, clay pottery, old swords, and other ancient artefacts. Because of this dream and their findings, these collectors felt charged by a mission to maintain the unity of the collection and never disperse it, in particular not to sell it. For many years, they kept the secret of their discoveries, but, as time passed, their mission became an unbearable burden. They finally decided to offer their land and the antiquities they had found to the royal family. To reach HRH Princess Sirindhorn, they advertised their project and, while waiting to be contacted by her, built a temporary museum on their land, the husband guarding it every night with a firearm. Without any official answer from the palace, they ultimately sold their collection to the authorities officially in charge, namely the Fine Arts Department (FAD), who took it to the Chumphon National Museum for evaluation.

When individuals seek and find ancient artefacts and offer to donate them to the Thai royalty, they mobilize both intense emotions and affects, and also patrimonialization issues, suggesting another facet of gems in a Buddhist monarchy, where the political centre is supposed to attract people, goods, and energetic vibrations. Such devoted collectors also bypass the law twice. In Thailand, since the fall of the absolute monarchy in 1932 and the subsequent bureaucratization of cultural management through the establishment of the National Institute of Culture (in 1942) and, later, a dedicated ministry (Peleggi, 2007), the royal family was restricted from intervening directly in Thai politics of culture. When ancient artefacts are unearthed, the discoveries are supposed to enter the national patrimony and, as such, are treated as state property. But the state officials in charge are often the subject of controversies and suspicions regarding antiquities management, portable antiquities in particular. As a consequence, many discoverers do not trust state authorities, preferring to donate their collections to the royal family, as in the case of the couple just mentioned, or to keep them for themselves, as in examples discussed further on. There are two main reasons for this mistrust. First, some individuals point out that these inalienable artefacts often disappear from the public eye after being taken by state authorities, hidden away in the storerooms of museums. Even those artefacts that are eventually displayed in permanent exhibitions throughout the country are not considered by specialists to be the most spectacular, rare, or historically relevant. Second, many people then ask the question, What happens to these most important remains? Some say they are protected in banks or treasury rooms. Others relate stories of antiquities disappearing from museum collections to the benefit of private collectors.

In the past few years, suspicions around illegal trafficking involving museum staff and private collectors have made the headlines of local and national newspapers. In 2011, for example, stories circulated about the trafficking of prehistoric beads from the Thalang National Museum in Phuket. In the Thai-Malay peninsula, in the absence of textual evidence and in a tropical environment, beads formed of glass, shells, metal (gold in particular), or hard stones (carnelian, rock crystal, garnet, amethyst) constitute among the best-preserved archaeological evidence of the early periods of the maritime silk roads (Bellina, 2007). They attest, in particular, to the existence from the first millennium BC of a network of related ports in this region, some being the first "cosmopolitan" urban establishments to develop along the silk sea routes in Southeast Asia (Bellina, 2007; 2017). Describing "Phuket's ancient bead saga," a Thai archaeologist from the FAD regional office noted in the columns of the *Phuket News* that "trading archaeological artifacts is a chronic problem [in Thailand] ... [M]any [Thai national artifacts] are in the hands of private owners" due to their "high value," but "it's impossible to keep an eye on every historical piece found in Thailand" (Phuket News, 2011). Illegal collecting of beads and other antiquities is not new in the country (Lertcharnrit & Carter, 2010; Marwick et al., 2013; Shoocongdej, 2011). Devastatingly for archaeologists, recent years have seen a significant increase in looting along the Thai-Malay peninsula, in the region of Chumphon in particular, where archaeologists are challenged by bead-collecting amateurs. When identified, ancient settlements provoke spectacular "bead rushes" (Bangkok Post, 2012), which not only feed networks of big collectors and a burgeoning national black market but also enable new uses and emerging practices among local inhabitants. These practices are related to aesthetics, leisure, and pleasure, and also to changing understandings of local heritage and the value of its preservation, which are enabling a recentred writing of history on locality.

In Sawi, Chumphon province, a group of ten men, inspired by a passion for old beads that might be worn, began to visit local caves that had been used as burial sites during prehistory. These collectors wear their beads on a daily basis. By keeping the beads close to the skin, they claim that they develop strong intimate relationships with the beads, arousing personal sensations and emotions. As such, the collectors exemplify the sensual attachments to minerals, charged not only with their vital components but also with personal experiences and emotions. While it was ethnographically impossible to reach royalty, conducting ethnographic fieldwork with these bead-collecting amateurs has offered different insights from those gleaned during similar work with the artisans and merchants mentioned previously, especially with regard to how some minerals are valued locally and might affect people through their possession and use. One member of the team explained, for example, that he never spent a day without grasping the beads of his necklaces. He observes them in attentive contemplation and speaks about them every day to his friends and relatives. Omnipresent, the beads are always at a reasonable distance from his hands and eyes, his physical contact with them becoming a vital necessity. While he neglects the glass beads that he gave to children and women when he found them, he developed over the years a preference for hard stone beads, which, he says, bind him to the entrails of earth and to a temporality before humanity, bringing him closer to an immutable nature. His necklaces are made of dozen of agates, carnelians, amethysts, and crystals. He takes care of them with permanent devotion, trying to maintain them in the physical state of their discovery. Familiar with their presence, he knows each of them perfectly and follows their change of colour and texture as he manipulates them in different weather and light, echoing the legends around the last king of Burma, King Thibaw (Meylan, 2012: 274–82). It is said that, several times a day, the king opened the strongboxes containing his collection of rubies, which was kept in his apartments. He would carefully take the stones out of their silk cases, one after the other, and contemplate them in the light for hours, neglecting official affairs.[16]

Like his brother and friends in the same group, this collector of ancient beads didn't assemble his necklace randomly but, rather, in keeping with his personal taste, following an original pattern inspired by an intact necklace found placed on a skeleton in a cave. To match this model, beads have to be paired according to various criteria (material, colour, size, shape, and also sharp edges, origin, and so on) around a central pendant, usually a bigger bead or a contemporary holy figure, such as a Buddha or a local saint endowed with potency, well known in Thai popular culture and Buddhism (Tambiah, 1984; Hemmet, 1989; Kitiarsa, 2012). To complete their necklaces, members of the team exchange beads among themselves and frequently discuss their arrangements. These beads circulate among wearers, so the necklaces are never static and keep evolving as finds of recent expeditions become available; new discoveries lead to new pairings, but always within the limits fixed by the original pattern. While these beads are exchanged among members of the team, all are able to recognize the beads they had given to and received from others. The team members endlessly comment on these beads' qualities, their trajectories, and the social relations they mobilize. As such, the necklaces, and the search for, care of, and exchange of beads, trace local networks among friends and neighbours that reflect and attest to a strong male sociability and a shared experience of local territory.

These beads also contribute to collectors' identification and sense of belonging as locals, indigenous to the region. The collectors' attachment to the local region is emphasized by the fact that they wear only beads from Chumphon province, whether they have uncovered the beads themselves or received them from friends and relatives. As inhabitants of the locality in which the beads were found, these individuals consider themselves to be legitimate collectors of the beads, well positioned to protect and produce knowledge about them. Although the team members had settled in the area only, at most, two generations ago, attracted by empty lands that are nowadays used for plantations, they claim a certain "autochthony" that, to them, justifies their rights over these ancient artefacts. Through their collective practices, they feel legitimized to claim an ancestral connection to what they consider their territory and heritage. They explained, for example, that they actively seek new sites to ensure that the beads and other antiquities found in these places can be kept within their own network and community. Presenting themselves as bead keepers, the team members are acting, they say, to prevent the beads from being captured by external groups of collectors who would disperse them into national and international markets; they are especially concerned that artefacts might end up in the collections of private collectors, mostly based in Bangkok. These individuals also challenge the work of archaeologists. Very suspicious, they accuse archaeologists of working with various networks of collectors (some based in other regions of Thailand) and of sending antiquities considered significant to the region's history away to national museums, where the artefacts are not just neglected but kept hidden from the public in closed vaults. In their view, these local artefacts are (mis)used by the state and its allies to reinforce a particular national narrative in a way that simultaneously makes them inaccessible to people in the regions from which they come.

In Thailand, beads and other evidence of prehistoric connectivity found in the Upper South remain at the margins of official historiography. The national narrative promoted in school textbooks and commemorated by monuments emphasizes historical periods related to Buddhist "Siamese kingdoms" of the central region from the thirteenth century onward (Sukhothai, Ayutthaya, Thonburi, and Bangkok). Until the 1960s, the hegemonic position achieved by the "royalist-nationalist" school over Thai historiography delineated a centralized, unified, ethnically and culturally homogeneous "Thai Nation" (Thongchai, 1994; Jory, 2003). Criticized from the 1970s onward with mixed success (Glover, 2006; Jory, 2011; Peleggi, 2016), these nationalistic views on the history of "Thai-land" – "the country of the Thai" – and "Thai-ness" are still largely spread around the country, reinforced by popular depictions in television dramas, for example. When the collectors of ancient beads from Chumphon capture local artefacts for their own use, they challenge both this Thai-centric official history and the state's heritage management policies over antiquities. In claiming a patrimonial status and favouring their local narratives and personal practices, sensations, and emotions over national patrimonialization and national history, they offer an alternative perspective from the geographical, social, and political margins on the management of antiquities.

Under the patronage of a local entrepreneur with whom they are related, members of this team have proposed handing their collection over to a heritage centre that their patron intends to build in the region. To conduct this project, they agreed to collaborate with professional archaeologists, who acted as intermediaries with authorities in charge of national heritage issues. Curious about history, the team members read archaeological books and engaged in discussions with scientists about ancient sites they have found and properties of the artefacts discovered, thereby adding their vernacular knowledge to the different "regimes of historicity" (Hartog, 2003) existing in the region. Unlike archaeological "plausible scenarios" (Grimaud, 2013) for the past history of the region under consideration, their histories are manifested in territorial claims, a risky experience of landscape valorised as an immersion in a wilderness (see chapter three of this volume) supposedly directly related to the past, and a daily sensorial experience of the ancient artefacts they wear. Through the recognition that will come with the patrimonialization of their collections, they expect to be acknowledged as experts on beads and to assert their legitimacy as "guardians" of the local heritage against "outsiders." This narrative stands in stark contrast to situations considered prevalent in various studies around the world, where local collectors, generally labelled "looters," are commonly presented as lacking an agenda beyond immediate economic motivations (Matsuda, 2005) and as being entirely in the hands of middlemen motivated only by potential profit.

Some members of the team have also come to feel a strong link to the local prehistoric people who fabricated and first wore these beads long ago, and who sometimes visit them in dreams leading them to the discovery of their treasures. This link to humans of the past, which the team members have developed while uncovering skeletal remains in the ancient people's last "home," is very physical, sensorial, and emotional. To collect beads, the team members climb mineral peaks, clear patches of creepers and trees, explore rock cavities, dig the soil with their hands, confront bats and ticks, and, when they are "lucky," handle bones and teeth that some collect after having the items blessed by the abbot of the Buddhist temple next to their home. To do so, the team members had to emancipate themselves from ancient local beliefs and practices partly related to Buddhism. Indeed, the sites they seek were once considered so potent and, as a consequence, so dangerous that these places were kept intact and protected by previous generations, the caves closed and worshipped only from the outside. When ancient remains were found on their lands, people later donated them to local Buddhist temples, where some of these artefacts remain today in curiosity cabinets. Through daily physical contact with beads qualified as "potent" or "*saksit*" (Reynolds, 2005), the contemporary collectors of antiquities embody the potency associated with their experiences of wilderness at the frontier of the cultural and historical world, and of ancient artefacts that are commonly used as protective amulets. In Thailand, the perceived potency of beads and ancient remains to act on humans' lives is one of the reasons for the development of a national market for portable antiquities. While some people, archaeologists especially, regret the appropriation of beads and other artefacts for individual purposes (as well as the associated destruction of ancient sites), these new uses attest to the social vitality of these objects as they are desacralized, resacralized, and, ultimately, entangled in dense sociability, most notably through renewed sensual and embodied engagement with people.

Interestingly, while archaeologists and amateur bead collectors have different attitudes regarding what the ultimate fate of these beads and artefacts should be, all maintain attachments, developed through physical contact, to these minerals. During fieldwork with the Thai-French archaeological mission in the Thai-Malay peninsula, I witnessed archaeologists working to trace back the "*chaînes opératoires*" of the beads they encountered, meticulously studying the surfaces and depths in order to document all of the past technical operations required for shaping, polishing, or perforating the beads. Try as they might to avoid expressing their own subjectivity and emotions in the process, these archaeologists were sometimes surprised by singular pieces that pulled them out of standard procedures, appealing to their eyes and hands more than expected, even though they may not have intended to engage emotionally with the materials over which they claimed and developed expertise. Bead keepers, on the other hand, were and are taken with beads through different processes, bringing subversive local appropriations and narratives of the past, and more often confrontation and contestation, into dialogue with conventional archaeological history as well as with official narratives and central authority, incarnated in governmental policies and agents or in royalty. In these geographical, historical, and social margins, the bead keepers preserve a close contact with beads belonging to the nation and, in a way, claim the exclusive right to manipulate them as the king does with other precious gems, including the Emerald Buddha and royal treasures described in previous sections. Unlike the royal family, however, bead keepers do not radiate this powerful intimacy and the potential of its effects at the scale of the nation, but keep it, intentionally, close to their own bodies, immediate social networks, and locality.

Conclusion

Minerals materialize global economic, political, and representational issues. Seized into dense meshworks of practitioners involving craftspeople, sellers, collectors, and others, they are turned into remarkable gems, both through narratives and their physical manipulation. In discussing three cases from ongoing research in Thailand, a renowned hub for gemstone production, exchange, and consumption, this chapter has offered a preliminary and comparative overview of the modalities by which jade, diamond, corundum, carnelian, agate, or amethyst become valued differently but, ultimately, each as precious in their own way. Connections to the iconic Thai kingship run through all three cases, revealing, in the first two especially, how singular minerals legitimatize its authority, the figure of the king in particular, and how, in becoming so closely associated with these figures of authority and their bodies, certain minerals gain an aura of uniqueness and preciousness. While these minerals become precious through rich storytelling that may involve generic accounts of their "potency," "naturalness," or "antiquity," or through individual stories in the biographies of certain named specimens, I have shown that their preciousness is not only a product of semiotics. Besides their entanglement in systems of representations, these minerals share a common quality: they enchant (Gell, 1994) those drawn to them with affecting properties that tend always to escape human efforts at normalizing them, whether through official royal historiographies, the standards of international markets, or archaeological analysis and typologies. Their qualities are, indeed, worked through three different processes: ritual and paraphernarial *accumulation* in the first case, material *purification* in the second, and territorial and temporal *sedimentation* in the third. While these minerals create uncertainties, they reveal themselves in the end as more versatile than their crystallization may lead us to suppose. By keeping and sometimes vigorously defending a close and exclusive contact with these minerals and manipulating them in various ways, people attempt to stabilize the qualities they find most precious so that they might better characterize, identify, and value them. Doing so requires close contact and ongoing physical engagement, the very substance of minerals repeatedly putting sensorial expertise and a certain degree of attention and emotion to the test, with standards being redefined, contested, or reinforced in the process. While the value of these and other precious minerals certainly emerges from their potential as objects of circulation and exchange (see chapter three of this volume), I have shown that their preciousness emerges even more obviously from the intense intimacies, pleasures, and emotions that they bring to humans through close contact, far from the pristine stage valued elsewhere (see the introduction to part two and chapter five of this volume). As such, the ontological instabilities of minerals resonate and trouble the human meshworks that catch them up, making minerals particularly good substances with which to think about the complexity of human engagements in the world.

NOTES

1 The previous world record was for the Graff ruby sold in November 2014.
2 The *News from the Palace* presents the changing of the Emerald Buddha costume on 16 March 2014. See MGR Online VDO (2014, 16 March).
3 "The city of angels, great city, the residence of the Emerald Buddha, capital of the world endowed with nine precious gems, the happy city abounding in great royal palaces which resemble the heavenly abode wherein dwell the reincarnated gods, a city given by Indra and built by Vishnukarn" (Sternstein, 1982: 11).

4 King Rama IX has since died on 13 October 2016, and his son was proclaimed the new monarch of Thailand on 1 December 2016 under the title of King Rama X.

5 King Chulalongkorn, Rama V, described the ceremony in a famous 1888 text, *Royal Ceremonies of the Twelve Months* (*Phraratchaphithi sipsong duan*). See Chulalongkorn (1888); Lingat (1935: 26).

6 The astrological dates for the ritual ceremonies are the first waning moon of the fourth, eighth, and twelfth months of the lunar calendar.

7 "Tai" is used in reference to an ethnolinguistic family and "Thai" to the nation of Thailand (previously Siam).

8 "According to Buddhist mythology, when a great Cakkavatti king or Universal Monarch appears in the world, this gem Jewel, which normally resides on Mt. Vibul, comes to him along with six other great gem possessions and remains in his care until the very end of his reign" (Reynolds, 1978: 176).

9 For an example of stone shaped into Buddha images, see Tambiah (1982: 17). Tambiah evokes a black rock not far from the city of Ayutthaya, where the Buddha descended from the sky and sat accompanied by his disciples. The rock was turned into five identical Buddha statues and distributed among vassal kingdoms of the royal patron who ordered these images to be carved.

10 Most scholars suggest that the statue is made of jadeite or nephrite from current Myanmar or Northern Thailand, where green stones of the same kind are still found (Rod-Ari, 2010: 59). R. Lingat evokes, for example, the region of Nan in the northern part of Thailand, which is home to a variety of green quartz a bit lighter than the Emerald Buddha statue (1935: 14).

11 This interpretation had been reviewed since by various scholars (see, notably, Rod-Ari, 2010: 59).

12 This practice is evident during the New Year celebrations in Laos, when family members gather together and visit each temple of the ancient capital to worship statues, the palladium of Luang Phrabang – the Phra Bang – included, which are moved for the occasion into a lower position to be reachable even by humble commoners. It is not clear if this collective close worshipping of the Phra Bang statue is related to the renewal of Lao cults under the socialist regime or if it pre-existed under the monarchy.

13 The statue continues, nevertheless, to preside over royal ceremonies through white cotton threads that physically link it to the place of ritual performances (Lingat, 1935: 24).

14 The summer season costume's crown and other ornaments suggest attributes of the kingship. The rainy and cold season's costumes, meanwhile, evoke the monastic life, with the monastic robe of the former evoking the annual three-month retreat that every monk is expected to complete and

the shawl of the latter recalling the wandering life monks are meant to live the rest of the year.

15 The research for this section was conducted with the Thai-French archaeological mission in the Thai-Malay peninsula and, in particular, in a collaborative project with B. Bellina (French National Centre for Scientific Research; CNRS), its director, and O. Evrard (French National Research Institute for Development; IRD). See Vallard, Bellina, & Evrard, 2015.

16 The king's collection was huge. Since the early sixteen century and the discovery of Mogok's mines, where most of the prestigious rubies had been found, the largest and most beautiful stones were reserved for the crown. Some say that King Thibaw was so obsessed by his gems that his negligence of state affairs led to the end of his reign under British colonial rule and, ultimately, his heartbroken separation from his precious 98 carat ruby named "Nga Mauk," his favourite, which disappeared mysteriously during his forced exile in India.

REFERENCES

Askew, M. (2002). *Bangkok: Place, practice and representation*. London: Routledge.

Bangkok Post. (2012, 8 December). Bead rush reveals dark deals. *Bangkok Post*. Retrieved from https://www.bangkokpost.com/news/local/325127 /bead-rush-reveals-dark-deals

Barnes, O. (2016, 17 October). From the Unnamed Brown to the Golden Jubilee diamond. *Diamond Authority*. Retrieved from http://www .thediamondauthority.org/the-golden-jubilee-diamond/

Bellina, B. (2007). *Cultural exchange between India and Southeast Asia. Production and distribution of hard stone ornaments (VI c. BC–VI c. AD)*. Paris: MSH.

Bellina, B. (Ed.). (2017). *Khao Sam Kaeo. An early port-city between the Indian Ocean and the South China Sea*. Paris: EFEO.

Bennett, J. (2008). Matérialismes métalliques. *Rue Descartes*, *59*(1), 57–66. https://doi.org/10.3917/rdes.059.0057

Bessy, C., & Chateauraynaud, F. (1995). *Experts et faussaires: Pour une sociologie de la perception*. Paris: Métailié. (Postscript in English to the second edition: Being attentive to things: Pragmatic approaches to authenticity, Paris: Ed. Pétra, 2014).

Brazeal, B. (2017). Austerity, luxury and uncertainty in the Indian emerald trade. *Journal of Material Culture*, *22*(4): 437–52. https://doi.org/10.1177 /1359183517715809

Brown, R.L. (1998). The miraculous Buddha image: Portrait, god, or object? In R.H. Davis (Ed.), *Images, miracles, and authority in Asian religious traditions* (pp. 37–54). Boulder, CO: Westview Press.

Caillois, R. (1970). *The writing of stones*. Charlottesville: University Press of Virginia.

Calvão, F. (2015). Diamonds, machines and colours: Moving materials in ritual exchange. In A. Drazin & S. Küchler (Eds.), *The social life of materials: Studies in materials and society* (pp. 193–208). London: Bloomsbury.

Chiu, A.S.C. (2012). *The social and religious world of Northern Thai Buddha images: Art, lineage, power and place in Lan Na monastic chronicles (Tamnan)*. Doctoral dissertation in Art and Archaeology, London: SOAS.

Chulalongkorn, K. (1888). *Royal ceremonies of the twelve months* (*Phraratchaphithi sipsong duan*).

De Beers. (2008). Golden Jubilee. Retrieved from https://web.archive.org/web/20080613143146/http:/www.debeersgroup.com/en/About-diamonds/a-few-famous-diamonds/Golden-Jubilee/

Dharabhak, R. (n.d.). Contemporary mission of "Royal Goldsmith." *Ploy Petch Magazine*. Retrieved from http://www.guruthaiantiquejewelry.com/english/eng-about-us.html

Ferry, E.E. (2005). Geologies of power: Value transformations of mineral specimens from Guanajuato, Mexico. *American Ethnologist*, *32*(3), 420–36. https://doi.org/10.1525/ae.2005.32.3.420

Ferry, E.E. (2013). *Minerals, collecting, and value across the U.S.-Mexican border*. Bloomington: University of Indiana Press.

Gell, A. (1994). The technology of enchantment and the enchantment of technology. In J. Coote & A. Shelton (Eds.), *Anthropology, art, and aesthetic* (pp. 40–63). Oxford: Clarendon Press.

Gibson, J. (1979). *The ecological approach to visual perception*. Boston, MA: Houghton Mifflin.

Glover, I.C. (2006). Some national, regional and political uses of archaeology in East and Southeast Asia. In M.T. Stark (Ed.), *Archaeology of Asia* (pp. 17–36), Malden, MA: Blackwell Publishing.

Graham, M. (2013). Thai silk dot com: Authenticity, altruism, modernity and markets in the Thai silk industry. *Globalizations*, *10*(2), 211–30. https://doi.org/10.1080/14747731.2013.786224

Grimaud, E. (2013). Archéologie et ventriloquie. Jeux de chaises et de choses au bord d'une tranchée archéologique. *Gradhiva*, *18*, 200–33. https://doi.org/10.4000/gradhiva.2750

Hartog, F. (2003). *Régimes d'historicité*. Paris: Seuil.

Hemmet, C. (1989). La Thaïlande, le pays au million d'amulettes. *Objets et Mondes*, *26*, 49–64.

Hennion, A. (2004). Une sociologie des attachements. D'une sociologie de la culture à une pragmatique de l'amateur. *Sociétés*, *85*(3), 9–24. https://doi.org/10.3917/soc.085.0009

Hennion, A., Maisonneuve, S., & Gomart, E. (2000). *Figures de l'amateur*. Paris: La Documentation Française.

Hennion, A., & Teil, G. (2004). Le goût du vin: Pour une sociologie de l'attention. In V. Nahoum-Grappe & O. Vincent (Eds.), *Le goût des belles choses* (pp. 111–26). Paris: MSH.

Hughes, R. (1997). Judging quality: A connoisseur's guide. In *Ruby & sapphire* (Chapter 10). Retrieved from http://www.ruby-sapphire.com/r-s-bk -quality.htm

Hughes, R. (2001). Sapphire connoisseurship. Passion fruit. Lotus gemology. *The Guide, 20*(2, part 1), pp. 3–5, 15. Retrieved from http://lotusgemology.com /index.php/library/articles/150-passion-fruit-a-lover-s-guide-to-sapphire

Ingold, T. (2000). *The perception of the environment: Essays on livelihood, dwelling and skill*. London: Routledge.

Ingold, T. (2001). From the transmission of representations to the education of attention. In H. Whitehouse (Ed.), *The debated mind: Evolutionary psychology versus ethnography* (pp. 113–53). New York: Bloomsbury Academic.

Ingold, T. (2007). *Lines: A brief history*. New York: Routledge.

Jory, P. (2002). The Vessantara Jataka, Barami and the Bodhisatta-Kings. The origin and spread of a Thai concept of power. *Crossroads: An interdisciplinary Journal of Southeast Asian Studies, 16*(2), 36–78.

Jory, P. (2003). Problems in contemporary Thai nationalist historiography. *Kyoto Review of Southeast Asia*, 3. Retrieved from https://kyotoreview.org /issue-3-nations-and-stories/problems-in-contemporary-thai-nationalist -historiography/

Jory, P. (2011). Thai Historical Writing. In A. Schneider & D. Woolf (Eds.), *The Oxford history of historical writing*: Vol. 5. *1945 to the present* (pp. 539–58). Oxford: Oxford University Press.

Kitiarsa, P. (2012). *Mediums, monks, and amulets: Thai popular Buddhism today*. Washington, DC: University of Washington Press.

Knappett, C. (2004). The affordances of things: A post-Gibsonian perspective on the relationality of mind and matter. In E. DeMarrais, C. Gosden, & C. Renfrew (Eds.), *Rethinking materiality: The engagement of mind with the material world* (pp. 43–51). Cambridge: McDonald Institute for Archaeological.

Koizumi, J. (2009). The making of "Thai silk" as a national tradition. Kyoto: Kyoto Working Papers on Area Studies (CSEAS), 27.

Ladwig, P. (2000). Relics "representations" and power: Some remarks on stupas containing relics of the Buddha in Laos. *Tai Culture, 5*, 70–84.

Lertcharnrit, T., & Carter, A. (2010). Recent research on Iron Age stone and glass beads from Promtin Tai, Lopburi. *Muang Boran Journal, 36*(4), 53–68.

Lingat, R. (1935). Le culte du Bouddha d'émeraude. *Journal of the Siam Society*, *27*, 9–38. Retrieved from http://www.siamese-heritage.org/jsspdf/1931 /JSS_027_1c_Lingat_CulteDuBouddhaEmeraude.pdf

Marwick, B., Shoocongdej, R., Thongcharoenchaikit, C., Chaisuwan, B., Khowkhiew, C., & Kwak, S. (2013). Hierarchies of engagement and understanding: Community engagement during archaeological excavations at Khao Toh Chong Rockshelter, Krabi, Thailand. *Terra Australis*, *36*, 129–41. https://doi.org/10.22459/ta36.12.2013.09

Matsuda, D. (2005). Subsistence diggers. In K.F. Gibbon (Ed.), *Who owns the past? Cultural policy, cultural property, and the law* (pp. 254–65). New Brunswick, NJ: Rutgers University Press.

Meylan, V. (2012). *Trésors et légendes. Van Cleef & Arpels*. Paris: Editions Télémaque.

MGR Online VDO. (2014, 16 March). *News from the palace*. Retrieved 4 May 2016 from https://www.youtube.com/watch?v=98bsYKKrn-w

Mus, P. (1933). Cultes indiens et indigènes au Champa. *BEFEO*, *33*(1), 367–410. https://doi.org/10.3406/befeo.1933.4628

Naji, M. (2009a). Le fil de la pensée tisserande. "Affordances" de la matière et des corps dans le tissage. *Techniques & culture*, *52–53*, 68–89.

Naji, M. (2009b). La formation des féminités à travers le tissage dans le Sirwa (Maroc). In C. Rosselin & M.P. Julien (Eds.), *Le sujet contre les objets ... tout contre – Ethnographies de cultures matérielles* (pp. 243–63). Paris: Éditions du CTHS.

Naji, M., & Douny, L. (2009). Editorial. *Journal of material culture*, *14*(4), 411–32. https://doi.org/10.1177/1359183509346184

Narula, K.S. (1994). *Voyage of the emerald Buddha*. Kuala Lumpur/New York: Oxford University Press.

Norman, D.A. (1999). Affordances, conventions and design. *Interactions*, *6*(3), 38–43. https://doi.org/10.1145/301153.301168

Notton, C. (1928). *The chronicle of the Buddha Sihing*. Bangkok (Thailand), n.p.

Notton, C. (1933). *The chronicle of the Emerald Buddha*. Bangkok (Thailand), n.p.

Peleggi, M. (2002). *Lords of things: The fashioning of the Siamese monarchy's modern image*. Honolulu: University of Hawaii Press.

Peleggi, M. (2007). *Thailand. The worldly kingdom*. London: Reaktionbooks.

Peleggi, M. (2009, 30–1 October). Icons, antiquities and mnemonic sites: The multiple lives of Thai Buddha images. In Objet/Knowledge Symposium conducted at UC Berkeley, CA. Retrieved from https://www.academia .edu/309637/Icons_Antiquities_and_Mnemonic_Sites_The_Multiple_Lives _of_Thai_Buddha_Images

Peleggi, M. (2013). From Buddhist icons to national antiquities: Cultural nationalism and colonial knowledge in the making of Thailand's history of art. *Modern Asian Studies*, *47*(5), 1520–48. https://doi.org/10.1017 /s0026749x12000224

Peleggi, M. (2016). Excavating Southeast Asia's prehistory in the Cold War: American archaeology in neocolonial Thailand. *Journal of Social Archaeology*, *16*(1), 94–111. https://doi.org/10.1177/1469605315609441

Phuket News. (2011, 20 September). Phuket's ancient bead saga continues. *Phuket News*. Retrieved from https://www.thephuketnews.com/phuket-ancient-bead-saga-continues-26911.php

Reynolds, C.J. (2005). Power. In D.S. Lopez (Ed.), *Critical terms for the study of Buddhism* (pp. 211–28). Chicago: The University of Chicago Press.

Reynolds, F.E. (1969). Ritual and social hierarchy: An aspect of traditional religion in Buddhist Laos. *History of religions journal*, *9*(1), 78–89. https://doi.org/10.1086/462596

Reynolds, F.E. (1978). The holy emerald jewel: Some aspects of Buddhist symbolism and political legitimation in Thailand and Laos. In B.L. Smith (Ed.), *Religion and legitimation of power in Thailand, Laos, and Burma* (pp. 175–93). Chambersburg, PA: Anima Books.

Rod-Ari, M.N. (2010). *Visualizing merit: An art historical study of the Emerald Buddha and Wat Phra Keo*. (Doctoral dissertation). Retrieved from ProQuest Dissertations and Theses database (3431812).

Roeder, E. (1999). The origin and significance of the Emerald Buddha. *Explorations in Southeast Asian Studies*, *3*(1), 15–34.

Shoocongdej, R. (2011). Public anthropology in Thailand. In K. Okamura & A. Matsuda (Eds.), *New perspectives in global public archaeology* (pp. 95–111). New York: Springer.

Sotheby's. (2014, 12 May). E-catalogue. Retrieved from http://www.sothebys.com/en/auctions/ecatalogue/2014/magnificent-jewels-noble-jewels-ge1502/lot.502.html

Sternstein, L. (1982). *Portrait of Bangkok*. Bangkok: Bangkok Metropolitan Administration.

Swearer, D.K. (2004). *Becoming the Buddha: The ritual of image consecration in Thailand*. Princeton, NJ: Princeton University Press.

Swearer, D.K. (2010). *The Buddhist world of Southeast Asia* (2nd ed.). New York: State University of New York Press.

Tambiah, S.J. (1978). *World conqueror and world renouncer. A study of Buddhism and polity in Thailand against a historical background*. Cambridge: Cambridge University Press.

Tambiah, S.J. (1982). Famous Buddha images and the legitimation of kings: The case of the Sinhala Buddha (Phra Sihing) in Thailand. *Res. Anthropology and Aesthetics*, *4*, 5–19. https://doi.org/10.1086/resv4n1ms20166675

Tambiah, S.J. (1984). *The Buddhist saints of the forest and the cult of amulets*. Cambridge: Cambridge University Press.

Thongchai, W. (1994). *Siam mapped: A history of the geo-body of a nation*. Honolulu: Hawaii University Press.

Tolkowsky, G. (2001). A letter from Gabi Tolkowsky, master diamond cutter. Retrieved from http://famousdiamonds.tripod.com/tolkowskyletter.html

Vallard, A., Bellina, B., & Evrard, O. (2015). Cristalliser l'histoire: La seconde vie des perles préhistoriques en Thaïlande péninsulaire. *Artefact – Techniques, histoire et sciences humaines*, Hors-série 1, 189–203. Retrieved from https://hal-univ-paris10.archives-ouvertes.fr/hal-01529075/document

Walsh, A. (2010). The commodification of fetishes: Telling the difference between natural and synthetic sapphires. *American Ethnologist, 37*(1), 98–114. https://doi.org/10.1111/j.1548-1425.2010.01244.x

Werly, R., & Maquin, B. (2004). *Jewelry treasures of Thailand*. Bangkok: Amarin.

Woodhouse, L. (2012). Concubines with cameras: Royal Siamese consorts picturing femininity and ethnic difference in early 20th century Siam. *TAP – Trans Asia Photography Review*, 2/2. Retrieved from http://hdl.handle.net/2027/spo.7977573.0002.202

Woodward, H.W. (1997). The Emerald Buddha and Sihing Buddhas: Interpretations of their significance. In N. Eilenberg, M.C. Subhadradis Diskul, & R.L. Brown (Eds.), *Living in accord with the Dhamma* (pp. 335–42). Bangkok: Silpakorn University.

5 Transparent Minerals and Opaque Diamond Sources

FILIPE CALVÃO

Concerted efforts by state and corporate actors to promote and certify a more transparent and ethical diamond mining industry have stumbled upon the near impossibility of technically establishing a reliable origin "signature" for rough diamonds. As pure carbon molecules, the origin of diamonds is often said to be unidentifiable, but a complex relationship between the value of rough stones and polished jewels in the business of "ethical" mining remains. Unlike other precious gemstones, the natural origin of diamonds is often undisclosed and remains shrouded in uncertainty for traders, buyers, and investors alike, despite ever more prevalent calls to certify and trace the mining source in tandem with tightening controls imposed by the threat of synthetic or lab-grown diamonds. In this chapter, I examine how the source (of extraction) and a diamond's origin (defined by its natural properties) are invoked or concealed in expert evaluations of value in Angola's trading rooms and Swiss jewellery auctions, building from my research with the Angolan actors of the mining industry – diggers, geologists, and traders – towards international jewellery experts and investors in high-end auction rooms.

Calculating the quality and defining a price for what sociologist Lucien Karpik (2010) calls "singular, incommensurable products" raises the fundamental conundrum of what constitutes the value of rarity and preciousness in resources extracted from nature. The much-publicized Lesedi La Rona rough diamond, the largest rough diamond found in over 100 years, offers an exemplary illustration of this problem. Featuring 1,109 carats of clean, top-white colour, this stone – mined in Botswana in late 2015 – was auctioned off in London by Sotheby's in July 2016. Despite estimates ranging from US$70 to US$150 million, the stone failed to secure a buyer.[1] This diamond is peculiar, not only for its extraordinary size but also for having been put on sale to the public in its uncut form, with a clear path from mine to market. As such, it calls for examining the value of quality and natural origin in public moments of auction exchanges.

Moving from Angola's diamond trading rooms to Geneva's jewellery auction sales, this chapter seeks to bridge instances of mining and high-end and expert trading. In so doing, it suggests that the labour of opacity and transparency creates a symmetrical relationship between ignorance and expertise, the elusive origin of rough minerals and polished jewels, and the contentious value of conflict stones and ethical commodities. Despite the industry's attempts to ensure "clean" and "conflict-free" gemstones straight from the source, this framework suggests that practices of regulation, legibility, and accountability in the extractive industry should also attend to the opaque spaces of value creation in the jewellery and art world. In the case at hand, a transparent and ethical mining industry in Angola is irrevocably tied to how the natural origin of gemstones is fabricated, promoted, or occluded in Geneva's auction rooms – where record-breaking jewels are offered for sale annually – and further concealed in opaque institutions such as the "*ports francs*" (or "free" ports) of the city, where one of the "world's greatest art collections" (Bowley & Carvajal, 2016), to recall a *New York Times* headline, is hidden and effectively kept out of sight and control. More broadly, this problem is compounded by the expanding application of "blockchain" technology to digitally track all diamond transactions or growing pervasiveness of undisclosed lab-grown synthetic diamonds.[2] The question, thus, is how to assess the value of diamond gemstones at once by their transparency and opaque origins.

Angolan Diamonds and the Place of Nature

The problem of natural origin and source of extraction – and, as I will demonstrate, of revealing and concealing said origin – raises a number of important ethical concerns, and Angola offers a stimulating vantage point for its examination. More than a decade after the end of its civil war in 2002, Angola's diamond mines still typify for many the repressive, authoritarian, and militarized disregard for the law, where subjects are violently and routinely stripped of political rights and life itself (Marques, 2011). And yet, the 1998 United Nations Security Council resolution banning Angolan "blood diamonds" (United Nations, 1998) spearheaded a new tendency towards transparency in the global mining industry. The country's billion-dollar diamond mining industry, centred in the Lunda Norte province, ranks as one of the top five diamond producers in the world; in 2015, Angola presided over the Kimberley Certification Scheme (Kimberley Process hereafter), an intragovernmental certification agreement meant to ensure transparent revenues in the diamond industry. Despite such championing of the cause of clean and legitimate diamonds, a social economy of opacity still saturates the work of extracting and trading diamonds in and from Angola. As I suggest here, the source of diamonds and other mineral properties matters for framing the interconnected problem of the nature of minerals, the making of mines, and the accumulation (and reproduction) of capital, thus connecting trading rooms in Angola to auction rooms in Geneva.

Let us consider this problem in the context of mining operations in Angola. In the diamond-rich province of Lunda, diggers are often described as the "best geologists" and have a strong reputation for spotting the most abundant diamond sources and richest *cibulo*, the colloquial name for the source of artisanal diamonds. This reputation often extends into the sale of diamonds, when petty traffickers call diggers into the trading room to "be more convincing." As one well-established diamond buyer put it, as if sharing a trick of the trade, one should look at the sellers' fingernails to make sure they look dirty enough to come from the *cibulo*. For this diamond merchant, there are areas within mining concessions that have not been sufficiently mapped: "[I]t is the digger that invests there by making his in-depth probe. That way, mines immediately know that the area has a good ore-content. The digger makes the prospection but he never gets anything from it." Given the likelihood of finding larger stones in alluvial mining,[3] the question is less about the sheer quantity of diamonds mined in artisanal operations but – as a part-time digger and miner put it – of striking that one diamond that "big buyers are greedy to see on their table" and "big rich people, big monopolies, guys with money, like the Queen Elizabeth, want to buy." But, outside the performative value of place (as indexed by diggers' fingernails), how does the source of a diamond truly matter?

When a trader dealing with rough diamonds intentionally depreciates a stone's quality by highlighting its natural imperfections, he may be inadvertently bringing nature back to the negotiation table. In some instances, moreover, the dealer actively seeks to challenge diggers' claims of the diamonds' source by turning the origin into a contentious subject throughout negotiations, with frequent references to the site of digging. In different moments, diggers invoke their expertise by inviting the trader to visit the mining site, saying, "One day I'll take you there, so that you can see how our work is." The point here is that, despite the plethora of technical instruments, such as electronic weight balances, colour grading machines, calculators, and miniature shovels, the very material being traded shifts as a result of the performative manipulation of the knowledge about nature and its objects (Calvão, 2015). What is being qualified and exchanged depends fundamentally on how each of the participants in this negotiation inhabits the place accorded to nature in a broader field of social and power relations (see chapter two, this volume, for more on the interconnection between the traded material and mining landscapes in the context of artisanal production).

Put differently, the source of a diamond is often a controversial rhetorical device, in-between references to the external properties of capital and global markets and what is ultimately present at the negotiation: the impure, dirt-filled, and often tarnished rough stone dug up from the earth. Against expert claims of internal charcoal residues, diggers argue from the standpoint of their authority vis-à-vis their immediate proximity to the earth-tainted surfaces of nature's diamonds. In response to a trader's low-priced offer, for example, a digger once asked in angered frustration: "How many holes do I need to open for a *kibula* [large stone] to come out? Tell me, how many holes?" In other words, the trader's expert evaluation of a diamond's origin – in which hidden imperfections inside the stone may be revealed – is challenged by the order of value signified by the actual protagonists of geological knowledge and the authority of those who map and dig up the source of diamonds. This situation is not unlike the scrapping community examined by Bell in chapter one, in which a form of expertise is performed in-between "gettin' dirty" and the systematic recycling of discarded electronic devices. Here, too, the opacity of these "unseen" components of the digital age (Smith, 2011) can be contrasted to the moral project of making minerals "clean" and transparent.

In contrast to the value of diamonds' source in Angola, little is usually known about the natural origin of diamonds beyond basic guarantees given to ethically and socially aware consumers. These may include the conflict-free certificate supplied by the Kimberley Process or blanket sourcing practices by large diamond traders and manufacturers, which tend to exclude African sources to privilege other mining areas deemed more "ethical."[4] Despite recent efforts to render this information more transparent, the sale of polished diamonds seems to build on an active withholding of knowledge. Aside from the assurances provided by certifications of ethical, responsible, and conflict-free diamonds, knowledge about the origin of diamonds remains, for the most part, shrouded in mystery, as we will see next.

From Mines to Markets

> Under normal circumstances, the value of a diamond can be defined by its carat weight, cut, colour, and clarity grade. But on rare occasions, a diamond will show something beyond, something special, an extraordinary charm that cannot be explained by words. This is when you know you have a legendary diamond in your hands. (Christie's, 2014: 256)

How exactly do traders, high-end jewellery buyers, and experts grapple with the contentious source of diamond stones in auction sales? Auction exchanges have been the subject of a well-established literature, mainly in economic sociology (Smith, 1989; Pardo-Guerra, 2013; Velthius, 2005). Despite the more recent engagement of anthropologists with auction settings and the problem of valorizing priceless commodities or defining market prices (for example, Geismar, 2013; Ferry, 2016), this literature joins studies of global market institutions and the socially embedded, semiotic, and performative financial practices that constitute them (Lipuma, 2017; Sassen, 2006; Zaloom, 2006; Zelizer, 2011). Here, I extend my enquiry of Angola's diamond trading rooms into two different domains: first, by including elements from high-end jewellery auctions organized by Sotheby's and Christie's in Geneva, Switzerland. These "magnificent jewels auctions," hosted by the two leading auction houses in the world, provide an important contrast and point of comparison with trading events in Angola, not only in terms of the magnitude of value dealt with but also in the very material nature of the object being traded. Second, I shift the locus of my enquiry from the mutual valuation of diamonds – framed in language use between market appraisal and the authority generated by technical expertise – to the problem of the source and natural origin of diamonds.

In attending Christie's and Sotheby's biannual auctions in Geneva – two of the most important in the luxury jewellery market calendar – I soon realized that these trading situations did not fall far from my previous research in Angola.[5] Despite the auctioneers' orchestrated gestures, there was an energetic and frenzied atmosphere in these auction rooms that resonated with Angola's packed trading rooms. With each increase in bids, for example, the audience's gaze riveted onto the exchange. Significantly, it was not merely the highest bids that were applauded by the audience but rather the intensity of bidding finales – irrespective of final sale price – that were received with clapping approval by the participants in both Sotheby's and Christie's auction sales. Conversely, when paddles were slow (or low) to bid, an auctioneer often acted disinterested and bored, physically leaning over the podium with a raised eyebrow, his aloof bodily demeanour visibly reminding the audience that they were not acting as was expected of them. In this, auctioneers performed in ways similar to Angolan diamond dealers, who can behave unfazed by the demands of a group of diggers or by the stones placed in front of them, negligently tossing stones around as if unmoved by their material properties. A game-like quality also pervaded these events, such as when the auctioneer alternated between joking with the audience – if the pace of the auction called for this sort of engagement – and fostering a competitive streak between bidders, not unlike Lunda's trading rooms.

The 2013 sale of the iconic Pink Star in Geneva – the most expensive piece of jewellery ever sold – offers perhaps a stark example of the conjunction of play, humour, and commotion that can also be observed in Angola's trading rooms. On that occasion, a tightly packed room at the Beau Rivage Hotel in Geneva awaited lot 372, the "largest known fancy vivid pink internally flawless diamond," at 59.6 carats, and "one of the world's greatest treasures," according to the auction's catalogue. Despite much anticipation and speculation, estimates were only available upon request prior to the sale. Ten years prior, when the stone made its first public appearance after being reduced from a 132.5 carat rough, renowned jeweller Yair Shimansky explained that it was virtually impossible to place a price tag on the stone: "How much would you pay for the Mona Lisa?" To the jubilation of the crowd in attendance that evening, the auctioneer's gavel signalled a record-breaking auction price. After a prolonged dispute with bidders on the phone, Isaac Wolf – a New York dealer present in the room – secured the deal at US$83 million. A few months later, the consortium of investors he represented defaulted on the acquisition, forcing Sotheby's to acquire the stone from the original seller who remained anonymous. But none of that mattered that evening, as the façade of solemnity de rigueur in high-end auctions gave way to the more playful Pink Panther theme song blasting through the room's loud speakers.

In both Angola and Geneva, it would seem, trading in diamonds (or other jewels) entails a performative exchange of words, a familiarity among participants, and a momentary disruption of orderly behaviour (such as with the Pink Panther theme song or in frequent bursts of applause). And yet, if the performative spectacle of auctions recalls trading moments in Angola, the value attributed to the origin of the object on sale in both contexts seems to represent an important element of distinction. In fact, for a stone of this rarity, the exact origin of the Pink Star diamond was not disclosed and remains highly speculative: news agencies repeated the vague claim that the diamond had been mined in Africa, and the auction house publicly stated that "it had no information" on where the stone came from, with some suggesting it had been sourced in war-torn Angola. Could the origin of this diamond be irrelevant to its public and market valorization? Put differently, what traces of carbon – markers of natural origin – are left imprinted in a polished stone?

Before being sold in the marketplace, as Ferry aptly demonstrates in a study of minerals and value circulation across the United States–Mexico border, all traces of mining are erased from collectible minerals so as to present, in the words of dealers, a stone virtually "untouched by human hands" (2013: 166). By forging this quality of "pristineness," Ferry argues, an active labour is put into rendering "the naturalness of mineral specimens visible and effective" (153). In the transition from rough to polished diamonds, I would suggest, a dissimilar intervention takes place. In fact, polished diamonds go through a process of "purification" by means of which all traces of nature are stripped from the surface of the stone and only the residues of human intervention remain: the jewel's provenance or the technical feat of perfecting light return to maximize its qualities of brilliance and transparency.

To be sure, the diamond industry tends to avoid an antonym for clarity, purposefully left uncoded to avert the assumption that a diamond could somehow be impure or polluted. Some internal occlusions, however, unavoidably remain after polishing and come to stand as reminders of the "dirtiness" of natural traces, pressed upon by the "fingerprints and 'hallmarks' of [nature's] genuineness" (Bruton, 1978: 133). Similarly, at the point when corporate and state actors of the diamond industry championed the Kimberley Process and the cause of a "clean," "transparent," and "accountable" diamond business, it has become evident that a certification scheme does not preclude untraceable paths of circulation or remove all uncertainty regarding diamonds' origins. In fact, these attempts at certifying the origin of diamond stones could be regarded, rather cynically, as an attempt at "cleansing" stones by moving them into a "legitimate" market.

It is not clear if and how the knowledge of a jewel's natural source would represent an added value to the provenance of "magnificent jewels," but here I wish to argue against the common assertion heard among diamond experts that the natural origin of diamonds is irrelevant for the mining industry. At least since the 1920s, in fact, the very same problem was a consistent concern for diamond companies in Angola, with a number of geological studies and reports addressing the possibility of tracing back diamonds (usually stolen) to their place of origin.[6] Conversely, much like traders in Madagascar speculate about the future uses of sapphires (Walsh, 2012) and Kivu miners are unaware of what happens to coltan (Smith, 2011), diamond diggers similarly puzzle about the use given to diamonds.[7] Unlike the curiosity manifest by diggers and miners, however, there seems to be an active ignorance in auction exchanges of the origins of the jewels up for sale. This deliberate indifference is not to suggest, however, that a mining source is irrelevant: a glance through the representatives of the major actors in the jewellery business present at the Basel Trade Show (Baselworld), one of the main events in the circuit of luxury jewellery, demonstrates the importance attributed to the public awareness of high-quality, ethically sourced stones.[8] Most companies and manufacturers claim to acquire rough diamonds from the "best" and most "reliable" sources, although only in one case, from the Zimmi region in Sierra Leone, is the actual origin of diamonds mentioned. If most dealers in emeralds, sapphires, or rubies take pride in acquiring their gemstones "directly from the source,"[9] whose value can then be seen as appreciating according to specific place of origin, most diamond buyers and participants in auctions seem to replicate the ignorance that miners and other producers demonstrate as to the intended use of the stones they unearth and trade. They often reiterate the common claim that it is impossible to "prove" the origin of an internally flawless diamond with no inclusion that could confirm its country of origin.

While origin may indeed be impossible to prove for the purest colourless diamonds, the same may not be true of coloured diamonds. According to a recent study led by the Gemological Institute of America (GIA) experts on colour saturation, fluorescence reactions, and concentrations of boron and nitrogen impurities in coloured diamonds, "one variable was largely excluded: geographic origin. Among the seventy-six stones studied, only a dozen were of known origin. This is not surprising, as information about a diamond's mine or even country of origin is rarely retained" (Gaillou, Post, Byrne, & Butler, 2014). There is, thus, a technical ability to certify and authenticate natural origin, particularly true in the case of Type IIa and Type IIb diamonds (devoid of nitrogen impurities, or carrying specific concentrations of boron impurities, which lead the stone to absorb some of its colours – blue or pink diamonds, most notably).

Let us return to the Pink Star diamond, for example. For one of the "Earth's great natural treasures," as David Bennett, famous auctioneer and recently appointed chairman of Sotheby's International Jewelry, called it, the precise location of the diamond's source, somewhat surprisingly, is still shrouded in mystery and speculation.[10] When asked about the stone's origins, Bennett suggested that "it came from an extraordinary rough, which apparently had 132 carats," skipping its mining source altogether and highlighting instead the moment in 2003 when the diamond was first publicly unveiled in Monaco before being loaned to the Smithsonian Institution. However, it is still telling that the origin of this pink diamond was not revealed, even more so given the recent emphasis on accountability, transparency, and responsible sourcing of gemstones (including industry initiatives such as the Kimberley Process, Extractive Industries Transparency Initiative, or the Responsible Jewelry Council). Is it believable that De Beers did not know, or cared not to share, the source of one of the rarest stones ever found, even more so when its value was premised on showcasing its unique properties of natural origin? Why did it appear that revealing its exact source seemed to pose some risk to its valuation – particularly if recent trends in the diamond industry highlight the ethical, responsible sources of these gemstones?

This diamond could well evoke the anxieties surrounding the source of diamonds in postcolonial Africa and the mining industry more broadly. According to rumours that circulated at the time of the stone's public viewing, De Beers's "lips are zipped as to the exact origins of the stone. Experts speculate it could come from Pretoria or Kimberley – but maybe even Angolan and Congolese mining giant, JFPI Corporation" (Lazare, 2003). As suggested earlier, news agencies speculated that the diamond had been mined by "De Beers somewhere in Africa in 1999, according to Sotheby's, which said it had no information on the exact geographic origin" (Nebehay, 2013). It is also interesting to register the controversial acquisition of this diamond. For some experts I had a chance to consult, the Pink Star diamond consignor was a very strong dealer who "probably requested to be paid quickly after the auction." One jewellery expert and investment manager I talked to in the aftermath of the sale found it "strange" that the buyer was acting on behalf of investors, as "no one would start an investment with such a purchase," which is to say before diversifying a portfolio and spreading the risk with smaller gemstones. A different conspiracy theory has emerged in the aftermath of this botched sale, with some suggesting that the investors were "spooked by information in the public domain linking BSGR (Beny Steinmetz's mining group, the previous owner of the stone) with Israeli human rights violations, information that leads many people to believe BSGR diamonds are de-facto blood diamonds" (Clinton, 2014). Although unlikely, this case would be the first instance of a diamond deemed "conflict-laden" after being polished. After years hidden from public scrutiny and without any assurance regarding its natural origin, this pink diamond surfaced in April 2017 in a Sotheby's auction organized in Hong Kong, where it became the most expensive gem ever sold at US$72 million.

Revealing Opaque Origins

Historically, diamonds have been mined and traded under a cloud of secrecy. Today, the dubious legality under which the diamond industry still operates reveals some of the very pitfalls of the Kimberley Process. According to a mining expert I interviewed, "unless you trace the origin to its mine ... there's no way to prove the geographical origin of the diamond after it has been mined." In other words, the Kimberley Process may not have been established to trace the origin of a stone but rather to maintain exclusive and secretive business practices based on elusive moral and ethical grounds. Paradoxically, and to return to the problem of source and ethical mining, the industry works with the consensus that the origin of diamonds is ultimately untraceable given the very natural properties of the stone. While this assertion holds true in most low-carat, undistinguished mass of diamonds, there are means to authenticate a stone's origin. To control irregular diamond supplies, De Beers has even pioneered some high-tech systems of detection: either by examining a speck of dust found on a rough diamond stone, measuring the degree of impurities to trace it back to its likely alluvial or kimberlite origin; analysing isotopes from samples of particles in order to identify its "authentic" signature; or inscribing authenticity onto the rough itself with laser tags, despite cases of fake inscriptions (Hart, 2001: 195–8). Similar techniques of "provenance proof" and "paternity tests" have been developed for emeralds and coloured gemstones to enhance their traceability and the knowledge about mining sources.[11]

Aside from these technological innovations seeking to trace and assess natural origin, which are expected to expand with the growing prevalence of synthetic gemstones, the source of diamonds is not always absent, obscured, or concealed. Blue diamonds hailing from the Cullinan mine in South Africa or the Lesedi La Rona diamond mentioned earlier are two recent examples, and the Canadian mining industry has built its reputation on the ability to certify its mining sources. There are two more important exceptions to this process of erasing diamonds' natural origins. The first is when an old diamond from the Kollur mines in the Golconda region of India reaches the market (joining other reputed diamonds such as the Hope or the Koh-i-Noor diamonds). Given that this source has been exhausted for a few centuries, the origin in this case is often conflated with provenance or the history of a particular stone. According to a diamond expert and jewellery investor interviewed in 2014, "the only diamond that could have higher value according to its origin is the Golconda diamond," as he recalled being struck by the "special life" of a diamond thought to come from this historical diamond source.

The Grand Mazarin diamond, named after seventeenth century French Cardinal Mazarin, confirms this exception. This diamond was sold for over US$14 million in Christie's 2017 fall auction in Geneva, in great measure given its 350 years of "enthralling stories involving a legendary cardinal, four kings, four queens, two emperors, two empresses, a spectacular jewel robbery, a notorious auction and the greatest jewellers of France" (Christie's, 2017: 268). Despite its complex and, at times, opaque history, "there is one single fact that is completely incontrovertible, it is that the Grand Mazarin originated in the legendary Golconda diamond mines" (Christie's, 2017: 268). This record is sumptuously illustrated over forty-seven pages in the auction catalogue, fittingly including a map of the famous Golconda mines. In statements to the press after the sale, François Curiel, Christie's chairman for Europe and Asia, called it "the diamond with the most prestigious and historic provenance still to be in private hands," narrating how it passed through the hands of the Sun King himself and Emperor Napoleon. And yet, even the light-pink, 19.07 carat Grand Mazarin diamond could not help falling into obscurity after its last public exhibition in the Louvre museum in 1962, having "disappeared again for more than half a century" (308).

A second exception to the opacity of natural origins, tellingly, was also brought to light in the same auction described earlier. In this case, the largest D flawless diamond ever to be sold in auction was hailed as originating from Angola by De Grisogono, the Swiss-based company that acquired and polished the 404 carat rough diamond. Despite adorning the cover and including glossy images of the craftsmanship involved in finishing the jewel, the catalogue makes no mention of Angola as the original source of this diamond. Instead, the stone's natural origin is vaguely described as a "miracle at birth," a "sudden moment of discovery" after "millions of years beneath the surface of the earth." After the sale, when asked about the source of the diamond in an interview in the empty auction room, auctioneer François Curiel simply referred to its natural features, pleased with the "auction fever" that led to more than US$118 million in sales. And yet, it was widely reported that the original rough diamond was sourced in Angola, a fact that De Grisogono equally publicized in a separate brochure distributed for the audience in attendance. The Geneva-based jeweller has made the rounds over its polemical ties with the Angolan regime and, in particular, the former president's daughter, who acquired a majority stake in the company for US$100 million in 2012 through an opaque "shell" company.[12] Whether this example foregrounds a new approach to disclosing the source of diamonds in order to reinforce their "natural" origin in contrast to synthetic gemstones (Walsh, 2010) remains to be seen. But, given the infamous record of human rights abuses in Angola's diamond mines, this marketing and branding strategy seems to have backfired when a self-proclaimed gem explorer went on a seven-day hunger strike to demand that the US$33 million proceeds from this sale go to the construction of a hospital in Angola. In other words, the transparent revelation of the diamond's origin did not performatively translate the opacity of the company's capital structure into a practice of ethical sourcing.

Conclusion

This chapter does not seek to build a case for why determining the sources of diamonds is relevant – despite claims to the contrary by jewellery experts, all diamonds have to be found, mined, and transported from somewhere. Ironically, the point could be made that the more transparent a stone is – as a sign of clarity and purity – the more likely it is that its source remains occluded by the lack of intrusions and internal occlusions that could give away its origin. Despite the exceptions noted earlier, the argument could be made that obscuring the origin contributes to its allure – concealing the source to fetishize the relations that go into producing its value – in conjunction with the long-established secretive mystique of holding sales of high-end rough diamonds outside public scrutiny.[13] The exclusion of the source in particular would seem to confirm the standardized anxieties concerning African diamonds and tales of secrecy and illicit capital flows, smuggling, and speculation, but it certainly seems to reduce to a simple narrative what are far more complex dynamics. Not only that, but the recent D flawless, 163 carat diamond from Angola may seem to confirm a new approach to corporate branding in the heretofore opaque business of diamond trading.

Pure and flawless rough stones present no unique features and are in every respect identical to any other transparent diamond. And yet, aside from the inherent complexity or potential futility of tracing the natural origin of a diamond, the revelation of its source and origin matter: not only to understand how diamonds are found and retrieved but also to grasp the work that goes into occluding that information as well. That much was already hinted at in the earliest geological report written about Angolan diamond concentrations in 1921: "In view of the peculiar conditions of the diamond market, it would be plausible to assume that the south African diamond miners would not desire to propagate information or means whereby other deposits might be found or from which diamonds might be cheaply extracted. I therefore wonder if the information they publish in regard to origin of diamonds ... is enlightening or misleading" Farnham (1921/1922).

Irrespective of new attempts to render business practices more transparent, the dubious legality under which the diamond industry still operates today reveals the paradoxical value of diamonds' sources – at once certified and regulated, but also uncertain and speculative. As I have tried to suggest, the production of geophysical and geological knowledge entails a far more complex conundrum of place and practices, firmly grounded in the ensemble of actors of diamond extraction surrounding the precise location and value of natural wealth. In fact, growing concerns over the place of minerals in the mining industry seem to contradict the general vagueness predicated in the jewellery industry with regards to the natural origin of diamonds. Despite ideologically driven claims to absolute transparency or the secretive reputation of the mining industry, the origin of a diamond is ultimately co-produced in the practices, qualities, and regulation technologies linking owner and worker, miner and manager, producer and consumer.

NOTES

1 The jeweller Laurence Graff acquired the rough diamond in a private sale in 2017. After cutting it into smaller stones, the now renamed Graff Lesedi La Rona weighing 302 carats was unveiled in 2019 as the highest clarity diamond ever certified by the Gemological Institute of America (GIA) (Graff, 2019).

2 The expanding number of blockchain pilot projects in mineral supply chains and the rise in synthetic gemstone production represent two of the main transformations in the mining industry. Distributed ledger technologies and blockchain solutions have been proposed to address the problem of conflict minerals, ensure due diligence standards, and improve commodity traceability. Paramount among these would be Everledger's and De Beers's Tracr blockchain initiatives for diamonds, or the Swiss Gemological Institute's GemTrack and Gübelin's paternity test solutions for coloured gemstones, among others. (See Calvão, 2019 and Cartier, Ali, & Krzemnicki, 2018 for an examination of blockchain mining and an overview of traceability initiatives in the gem sector, respectively.) On the rise of uncertified lab-grown synthetic diamonds entering the market, the quarterly journal of the Gemological Institute of America,

Gems & Gemology, carries frequent reports of synthetic or natural imitation diamonds. See, for example, the case of a natural diamond with synthetic coating (Moe, Johnson, D'Haenens-Johansson, & Wang, 2017) or a synthetic gem with traces of a diamond signature (Pakhomova et al., 2018).

3 Treatment plants or diamond mills in most industrial mines are equipped with recovery grids that reject stones above a certain grade, given the unlikely event of finding diamonds above a certain number of carats. Inadvertently, as the head manager of one of these treatment plants explained to me, this practice may lead to missing abnormally large stones.

4 This much is made clear by Tiffany's refusal to import diamonds from Angola as well as other purchasing policies privately disclosed by established jewellers that favour Canadian sources to the detriment of Angolan or Congolese sources.

5 Calvão (2015) explores how the materiality of diamonds calibrates value exchanges in Angola's trading rooms. In Calvão (2017), I examine the forms of ritual intimacy and knowledge production surrounding the interpretation of a divinatory apparatus in colonial Angola, where the value of diamonds is locked in occult idioms of secrecy and surveillance.

6 See Farnham (1921/1922). Admittedly, this report says that "apparently very little" is "known about the origin of the diamond."

7 Despite this apparent general oblivion as to the terms linking both ends of the supply chain of mineral commodities, diamond traders in Lunda routinely invoke international diamantaires as a negotiation strategy and as a sign of their privileged knowledge of the market ("I know the market, you have to trust"). This tactic does not mean, however, that diamond traders are knowledgeable about the consumer market for diamonds. On separate occasions, I was asked by different traders how much diamonds were sold for in North America or Europe.

8 However, the 2015 and 2016 exhibition catalogues only mention the Kimberley Process when describing the gemstone dealer CIRCA's "ethical and trusted sources in non-conflict regions, in adherence with the Kimberley Process protocols" (Baselworld 2015: 135) and the diamond trader Vulcan Co's practices of "only buy[ing] conflict-free diamonds from reliable sources and stones that have Kimberley certification" (Baselworld 2016: 237).

9 For example, in the 2019 GemGenève gem and jewellery show in Geneva, I witnessed one exhibitor pitching his supply of rhodolite garnet to a putative customer as coming from a "special lot, directly from the mine" in Mozambique. Although the exhibitor was unable to identify the mine or its precise location, this garnet found in Mozambique would contain a stronger red colouration than other garnet sources according to this gem expert. Symptomatically, this trade fair included an afternoon panel

discussion with different experts dedicated to "The gemstone market: From traceability to the responsibility of its actors."

10 David Bennett in an interview with Vivienne Becker for Sotheby's YouTube channel. Retrieved from https://www.youtube.com/watch?v=Tt0CaP2eQbE

11 For example, the Swiss-based Gübelin Gem Lab has recently introduced an "emerald paternity test" to trace the "provenance of emeralds back to the exact mine," expanded in 2019 to integrate blockchain-based solutions for the traceability of coloured gemstones (Gübelin Gem Lab, 2019).

12 The acquisition was revealed in 2014 by different news outlets. Records unveiled at the time showed that the operation channelled funds from a state-owned Angolan company through a Maltese-registered shell company to fund the acquisition of the Swiss jeweller.

13 This much seems to hold true of the Lesedi La Rona diamond, as revealed by Laurence Graff in an interview with Matthew Hart for a *Vanity Fair* piece. According to Graff, renowned diamantaire and arguably the market's biggest mover and shaker, "we don't like it, what they're doing. It's just not how it's done. We don't want to have to expose ourselves in public [at an auction]. To contend in the open arena, we find it undesirable" (Hart, 2016).

REFERENCES

Baselworld. (2015). *Selection. Stones and pearls*. Exhibition catalogue. Hamburg: Untitled Verlag und Agentur GmbH & Co. KG.

Baselworld. (2016). *Selection. Stones and pearls*. Exhibition catalogue. Hamburg: Untitled Verlag und Agentur GmbH & Co. KG.

Bowley, G., & Carvajal, D. (2016, 28 May). One of the world's greatest art collections hides behind this fence. *The New York Times*. Retrieved from https://www.nytimes.com/2016/05/29/arts/design/one-of-the-worlds -greatest-art-collections-hides-behind-this-fence.html

Bruton, E. (1978). *Diamonds*. Radner, PA: Chilton Book Company. (Original work published 1970)

Calvão, F. (2015). Diamonds, machines and colours: Moving materials in ritual exchange. In A. Drazin & S. Küchler (Eds.), *The social life of materials: Studies in material and society* (pp. 193–208). London: Bloomsbury Academic.

Calvão, F. (2017). The company oracle: Corporate security and diviner-detectives in Angola's diamond mines. *Comparative Studies in Society and History*, 59(3), 574–99. https://doi.org/10.1017/S0010417517000172

Calvão, F. (2019). Crypto-miners: Digital labor and the power of blockchain technology. *Economic Anthropology*, 6(1), 123–34. https://doi.org/10.1002 /sea2.12136

Cartier, L., Ali, S., & Krzemnicki, M. (2018). Blockchain, chain of custody and trace elements: An overview of tracking and traceability opportunities in the gem industry. *Journal of Gemmology, 36*(3), 212–27. https://doi.org/10.15506/JoG.2018.36.3.212

Christie's. (2014, 14 May). *Magnificent jewels.* Auction catalogue. Geneva: Manson & Woods Ltd.

Christie's. (2017, 14 November). *Magnificent jewels.* Auction catalogue. Geneva: Manson & Woods Ltd.

Clinton, S. (2014, April 18). $83 million diamond default: Sotheby's and Israeli war crimes. *Memo.* Retrieved from https://www.middleeastmonitor.com/20140418-83-million-diamond-default-sothebys-and-israeli-war-crimes

Farnham, C.M. (1921/1922). Geological field notes, Angola, 1 May 1921 to 20 August 1922. Angola Historical Archive, Cx. 4871, Lunda.

Ferry, E. (2013). *Minerals, collecting, and value across the U.S.-Mexican border.* Bloomington: University of Indiana Press.

Ferry, E. (2016). Gold prices as material-social actors: The case of the London gold fix. *The Extractive Industries and Society, 3*(1), 82–5. https://doi.org/10.1016/j.exis.2015.11.005

Gaillou, E., Post, J.E., Byrne, K.S., & Butler, J.E. (2014). Study of the Blue Moon diamond. *Gems & Geology, 50*(4), 280–6. https://doi.org/10.5741/gems.50.4.280

Geismar, H. (2013). *Treasured possessions: Indigenous interventions into cultural and intellectual property.* Durham, NC: Duke University Press.

Graff. (2019). Introducing the Graff Lesedi La Rona. Retrieved from https://www.graff.com/famous-diamonds/lesedi-la-rona/

Gübelin Gem Lab. (2019). Provenance proof. Retrieved from https://www.gubelin.com/cms/en/gemmology/unique-expertise/provenance-proof

Hart, M. (2001). *Diamond: A journey to the heart of an obsession.* New York: Walker Publishing Company.

Hart, M. (2016, 5 August). Why buyers shunned the world's largest diamond. *Vanity Fair.* Retrieved from https://www.vanityfair.com/news/2016/08/why-buyers-shunned-the-worlds-largest-diamond

Karpik, L. (2010). *Valuing the unique: The economics of singularities.* Princeton, NJ: Princeton University Press.

Lazare. (2003). Mystery of the priceless pink diamond. *PriceScope* (Community forum). Retrieved from https://www.pricescope.com/community/threads/mystery-of-the-priceless-pink-diamond.7431/

Lipuma, E. (2017). *The social life of financial derivatives: Markets, risk, and time.* Durham, NC: Duke University Press.

Marques, R. (2011). *Diamantes de sangue: Corrupção e tortura em Angola.* Lisboa: Tinta-da-China.

Moe, K.S., Johnson, P., D'Haenens-Johansson, U., & Wang, W. (2017). A synthetic diamond overgrowth on a natural diamond. *Gems & Gemology, 53*(2), 237–9.

Nebehay, S. (2013, September 25). "Pink Star" diamond could fetch $60 million, Sotheby's says. *Reuters*. Retrieved from https://www.reuters.com/article/auction-diamond/pink-star-diamond-could-fetch-record-60-million-sothebys-says-idUSL5N0HL1T620130925

Pakhomova, V., Fedoseev, D., Kultenko, S., Karabtsov, A., Tishkina, V., Solyanik, V., & Kamynin, V. (2018). Synthetic moissanite coated with diamond film imitating rough diamond. *Gems & Geology, 54*(4), 460–2.

Pardo-Guerra, J.P. (2013). Priceless calculations: Reappraising the sociotechnical appendages of art. *European Societies, 15*(2), 196–211. https://doi.org/10.1080/14616696.2013.767926

Sassen, S. (2006). The embeddedness of electronic markets: The case of global capital markets. In K.K. Cetina & A. Preda (Eds.), *The sociology of financial markets* (pp. 17–36). New York: Oxford University Press.

Smith, C. (1989). *Auctions: The social construction of value*. New York: Free Press.

Smith, J.H. (2011). Tantalus in the digital age: Coltan ore, temporal dispossession, and "movement" in the Eastern Democratic Republic of the Congo. *American Ethnologist, 38*(1), 17–35. https://doi.org/10.1111/j.1548-1425.2010.01289.x

United Nations. (1998). Security Council resolution 1173. Retrieved from http://unscr.com/files/1998/01173.pdf

Velthius, O. (2005). *Talking prices: Symbolic meaning of prices on the market for contemporary art*. Durham, NC: Duke University Press.

Walsh, A. (2010). The commodification of fetishes: Telling the difference between natural and synthetic sapphires. *American Ethnologist, 37*(1), 98–114. https://doi.org/10.1111/j.1548-1425.2010.01244.x

Walsh, A. (2012). *Made in Madagascar: Sapphires, ecotourism, and the global bazaar*. Toronto: University of Toronto Press.

Zaloom, C. (2006). *Out of the pits: Traders and technology from Chicago to London*. Chicago, IL: The University of Chicago Press.

Zelizer, V.A. (2011). *Economic lives: How culture shapes the economy*. Princeton, NJ: Princeton University Press.

6 Gold, Ontological Difference, and Object Agency

LES W. FIELD

My companions and I suffer from a sickness that only gold can cure.

Hernán Cortés
(quoted in Bachmann, 2006: 162)

Oh, most excellent gold! Who has gold has a treasure [that] even helps souls to paradise.

Cristóbal Colón
(quoted in Bernstein, 2000: 1)

Quotes such as these routinely characterize a literature called the "history of gold," a genre written both for scholars and for wider audiences. Not infrequently, a new "history of gold" is published. Peter L. Bernstein's *The Power of Gold: The History of an Obsession* (2000) is an excellent example of a best-selling volume that garnered significant praise from the *Wall Street Journal*, the *Financial Times*, and other such authorities; more recently, Matthew Hart, who has also written about diamonds, released *Gold: The Race for the World's Most Seductive Metal* (2013), which was also a best-seller. An iconic and Wall Street–oriented volume, Timothy Green's *The World of Gold: The Inside Story of Who Mines, Who Markets, Who Buys Gold* was released in 1993. Large, attractive, lavishly illustrated coffee table books such as Hans-Gert Bachmann's *The Lure of Gold: An Artistic and Cultural History* (2006) also appear perhaps every decade, with C.H.V. Sutherland's *Gold: Its Beauty, Power, and Allure* (1959) setting an early post-war standard. There are many, many more. Well-known, authoritative, and much more, scholarly volumes also form part of the "history of gold" genre, while departing from many of the popular literature tropes. Pierre Vilar's *A History of Gold and Money, 1450–1920* (1976) suggests that the historic role of gold should not be seen as either singular or isolated from the histories of other key metallic commodities such as silver or copper. Similarly, the more recent *War and Gold: A 500-Year History of Empires, Adventures and Debt* by Kwasi Kwarteng (2014) deeply contextualizes gold within imperial histories of conquest and economic cycles. In this way, the scholarship focused on gold is distinct from mass-marketed and broadly accepted ideas about gold; at the same time, both popular and scholarly understandings of gold maintain its agentive character, and it is the assumed agency of gold that provokes the consideration I elaborate here.

The conceit of the "history of gold" literary genre relies upon two narration idioms. The first idiom is comprised of a series of uncontested and apparently incontestable assertions about the physical properties of gold that render it a unique substance. Moreover, these assertions self-evidently and commonsensically explain, by way of material immanence, why gold's role in history has been what it is, specifically with respect to gold's relationship with value and money, as well as to gold's role as a (Western) symbol for love, fidelity, and marriage. A second idiom, closely related to the first, directs an occidentalizing historical gaze towards the history of gold, describing its earliest development in relation to Western narratives about the beginnings of civilization in Egypt, Mesopotamia, and the Mediterranean. These idioms tell a story that reiterates and reifies the unity between the history of gold and the history of coinage and money *because* of gold's unique physical character, which consequently erases all non-Western uses, understandings, and elaborations of gold. Those non-Western understandings and uses of gold are axiomatically folded and subsumed into the "history," which is the Western history. Gold assumes an agentive persona in the "history of gold," as the quotations at the beginning of the chapter make plain. Gold became money because of gold's character; gold twists the minds of men; gold's economic role determines the historic fate of nations; gold shows love; gold shows commitment. In the "history of gold," gold's purposes flow out of gold's character; rather than humans using gold for certain purposes in history, gold plays a role in history. In these mass-marketed, dominant narratives, gold, as itself, is an actor in conjunction with human beings but also distinct from human beings.[1]

My intention in this chapter is twofold. First, I query the occidentalist "history of gold," which collapses gold's history into a narrative about gold, money, and Western tropes of emotion, by probing the subject of pre-Columbian gold, referenced in the quoted exchanges between Columbus and the Spanish monarchs at the beginning of this chapter, specifically using a literature about gold in *what is now*

Colombia (which I will hereafter abbreviate as WINC) and working with a discussion in contemporary anthropology about ontological difference. The prevailing occidentalizing histories of gold reduce the non-Western ontologies, uses, and histories of gold to roadkill, consumed by the dominant narrative on the highway of Western advance. The "discovery" of the "New World," and in particular the conquest of Mesoamerican and Andean civilizations, always stands out because of the sheer scale of larceny and (literal) liquidation of a set of alternative gold worlds at the hands of Europe's gold world. I enquire into that encounter because it is an enticing place to engage with a discursive field, au courant in anthropology, defined by debates about ontology (see Alberti et al., 2011). I focus upon the Andean civilizations, which existed (and in some corners still do to a certain extent) in WINC, and discuss the existence (past and/or present) of radically alternative ontologies of gold in Native American civilizations.

Second, on the basis of my discussion of ontological difference provoked by the Western "discovery" of the golden treasures of pre-Columbian civilizations, I address the agency of gold – gold as historical actor – using different approaches, including, on the one hand, a Marxian analysis concerned with agency and commodity fetishism and, on the other, variants of what has been called the "new materialism," which also address the agentive role of objects. As a scholar who has used Marxist theory for over three decades, my consideration of the agency of gold is grounded in the Marxist approach I will detail, whereas my use of the work of "new materialists" and also approaches from anthropologists of the "ontological turn" could be described as exploratory. The dominant view about gold – composed of multiple and complex mixtures of discourse originating on Wall Street and in the global finance marketplace, the coffee table book histories, and the popular press, not to mention the vociferous twenty-first century gold enthusiasts – already embraces the view that gold *is* an actor. Since that view is so accepted as to seem practically commonplace, an article of common sense, what do "new materialist" frameworks have to offer?

Both analytic exercises seek to undermine "the [dominant] history of gold" and the conditions of thought that make it possible. Gold, I am concerned to argue, is not a singular, stable "it," even though in both Western and Amerindian ontologies, for example, a sociocultural, political, and economic "there" exists, where gold made and still makes sense and is used. In conveying such an argument, perhaps conceptual, if not physical, space is created to speculate about the future(s) of gold.

Ontology and Difference

The most dramatic example of the selectively collapsed character of the occidentalist "history of gold" occurs in the treatment of the Spanish conquest of the New World, in particular the subjugation of the Inca Empire and the entire Andean region, including those lands that much later became Colombia, where I conduct fieldwork. Columbus's declaration to Ferdinand II and Isabella, cited at the beginning of this chapter, sets the stage for a tale of grand larceny and mass death, to be sure, but also one in which the historical character of pre-Columbian gold was terminated in rapid, sharp strokes. For popular authors such as Bernstein, the story is exclusively about Francisco Pizarro; the gold of the Incas is iterated as "goblets, ewers, salvers, vases in great variety, ornaments and utensils, tiles and plates, curious imitations of plants and animals, and a fountain that sent up a sparkling jet of gold ... golden diadems, earrings, bracelets, and plaques" (Bernstein, 2000: 129), not to mention the throne of the Inca emperor Atahualpa, all cited in a page and a half in the context of the profoundly material and metaphorical melting down of all of these items into bullion by Pizarro's men. Bernstein notes the dollar value of those gold bars, yet hesitates to entertain even brief mention of what cultural, ideological, material, religious, or any other kind of significance gold may have held for the Incas. Hart writes:

> The Spaniards [Atahualpa] perceived had an appetite for gold. He could not have fully understood it, because the Incas had no money. They valued gold for the way it could be worked. It had its place in the adornment of nobles and in sacred rites, but even in those functions gold was not the top material. (Hart, 2013: 36)

Like Bernstein, Hart presents the main narrative flow in reference to the destruction of the Incas' gold objects and their dollar value as gold. In Sutherland's rendering, history arrives in this hemisphere with the Spanish; six pages of narrative draw the line from Columbus to Cortés to Pizarro in the decimation and devastation of Mexico and Peru, in which the author focuses in just three paragraphs on describing what kind of gold stuff the Aztecs and Incas made. No mention is given to what that stuff meant to pre-Columbian peoples and societies. The lavish illustrations in the Bachmann book are accompanied by text that elaborates more detail about the goldworking technologies and techniques developed by pre-Columbian Mesoamerican and

Andean civilizations, with only a passing glance at what gold meant to these peoples:

> [T]he material value of raw gold in pre-Columbian cultures was scarcely more than that of clay or salt – for the Incas and their predecessors, jade was always far more precious than gold – but designed and shaped, it nevertheless became the symbol of the sun and its creative energy as well as of a divine, otherworldly power. At the same time it was a status symbol for tribal chiefs, priests and worthy warriors and was lavishly sacrificed to gods and ancestors. The Spaniards' lust for gold – which Cortés summed up by saying "My companions and I suffer from a sickness that only gold can cure" – was therefore utterly incomprehensible to them. (Bachmann, 2006: 162)

Fortunately, a succession of both archaeologists and ethnologists who work in the Andes have not been content with such formulaic dismissals of pre-Columbian gold, and I will rehearse some of their work as it pertains to WINC against the backdrop of what some are calling "ontological anthropology" (Bessire & Bond, 2014). While Bessire and Bond's recent critique of this turn expends much of its force upon the approach's speculative, futuristic, world-making ambitions, they also focus upon the manner in which ontological anthropology has been ill equipped to "account for actually existing Indigenous alterity" and, consequently, "artificially standardizes alterity itself" (445). It is the latter arm of critique in their insightful analysis that resonates with my backwards gaze at the Spaniards' arrival in the Western hemisphere some 500 years ago.

Before addressing the anthropological work that focuses on the ontological statuses of pre-Columbian gold in WINC, I entertain the possibility of human civilizations in which the ontological status of gold has very little significance. That possibility simply seems *impossible* in the dominant "histories of gold," even though gold's presence in human history goes back, as far as is currently known, not even seven thousand years,[2] whereas other materials such as shell (more than 100,000 years of human history), ivory (perhaps 40,000 years of usage), and amber (at least 13,000 years of usage) form part of grave assemblages, mortuary sites, and other human habitations for a much longer time. The rhetorical strategy of the dominant "histories of gold" is to shroud gold's relationship with humans in the mists of indeterminately deep historical time. Within the approximately 6,500-year time frame during which (some) humans began relationships with gold, what of examples of societies for which gold, even if known, was considered relatively

unimportant? In the memoir *Lame Deer, Seeker of Visions*, Lakota sha-
man Lame Deer purports that the Lakota had known about gold long
before white people ever came to the Black Hills, where a great deal of
gold was lying around on the surface and in streambeds (Lame Deer &
Erdoes, 1972). Lame Deer claims that the Lakota knew about gold but
it did not matter to them – one could not eat gold, one could not use it
to keep warm, one could not make useful tools or implements from
it, and, in addition, it seemed that its qualities exercised few, if any,
charms upon the Lakota.

One should obviously not take this testimonial as an innocent or
transparent representation of the pre-conquest Lakota world view.
Lame Deer's narrative not only represents a current within Lakota
political-nationalist discourse of the mid-twentieth century, which
explicitly resisted the depredations of the settler-colonial state; his philosophy was also a fellow traveller with 1960s and 1970s counterculture
discourses that rejected materialism and mainstream symbols of wealth
and power. Any ethnographic narrative about gold likely bears the impacts and markers of the encounter with European and perhaps other
forms of colonialism. The point of seeking narratives such as Lame
Deer's is to provoke an imaginary that could deny gold significance.
That would seem an important step in theorizing gold in a way that
denies it an a priori actor status, as I discuss in the next section, as well
as providing a different resonance with Bessire and Bond's critique of
the ontological anthropology turn:

> [Ontological anthropology] is premised on a story of the South American
> Primitive ... [I]t is premised on identifying an "Amerindian multinatu-
> ralist ontology" and describing it as the opposite of "modern, Western
> European," "mononaturalist-monoculturalist philosophy," and the bina-
> ries of nature-culture on which this "modern ontology" is based. (Bes-
> sire & Bond, 2014: 442)

Bessire and Bond elaborate the premise of a nature-culture binary
in Western ontology and the different combinations of mono- and
multi-naturalism and culturalism they detect as the boundary between
Western and Amerindian ontologies in the work of Eduardo Viveiros
de Castro (1998; 2013, for example) and Eduardo Kohn (2013), among
many others. In doing so, Bessire and Bond critique ontological anthropology's assumption that an "incommensurability of difference"
sharply demarcates and bounds Western from Amerindian ontologies. This binary, they argue, rather than ushering in a new kind of
anthropological analysis, instead reasserts one of anthropology's (not

to mention modernism's) most colonial assumptions about the primitive versus the modern. Bessire and Bond use their own ethnographic work (for Bessire among the Ayoreo of Paraguay, and for Bond in the worlds that hydrocarbons destroy and remake in the United States) to show that the ontological turn not only re-establishes old boundaries between Western and non-Western societies but also relies upon surprisingly ahistorical, even archaic, assumptions about the static cultural worlds of "primitives." The authors also see the ontological turn in the work of Viveiros de Castro (see also Skafish, 2013) as a strange suturing between the philosophical work of Gilles Deleuze (classically found in Deleuze & Guattari 1983; 1987) and Amerindian, particularly Amazonian, cosmology. That connection means that anthropology's most promising future is what Martin Holbraad (2012) called "comparative ontography," which Bessire and Bond (2014: 447) again reference to a speculative futurism that consigns Amerindian ontologies to self-contained boxes outside of the "West."

Like Bessire and Bond, I argue against the rigidly reified dualism between Western and Amerindian ontologies, primarily because such a view simply is inadequate, if not tangential or even inimical, to ethnographic work with Indigenous peoples in the present and to the multiply complex and often contradictory mixtures of accommodation with and resistance to colonialism and nation-states that characterize Indigenous lifeways and Indigenous alterity. Such complexities emerged at the very beginning of the "encounter" between Amerindian peoples and Europeans at the end of the fifteenth century. The observation that Indigenous ontologies can never be innocent of the impacts of colonialism need not impede investigation of, indeed fascination with, difference and the contrast between what are still, nevertheless, considered by ontological anthropologists to be substantially different "worlds." In my work with a constellation of Native American tribes, scholars, and leaders in California (Field, 2008), I engaged in many conversations that contrasted Native and non-Native ontologies. What non-Natives, such as anthropologists, denominate "ritual regalia" are sentient beings "with agency and destinies linked to yet distinct from the humans with whom they cohabitate" (39). In De la Cadena's (2015) recent book, she delves deeply into the incommensurabilities that distinguish Native and non-Native ontologies, anthropology's incapacity to come to terms with Indigenous ontology because of the domination of the culturalist paradigm, on the one hand, and the limitations of translation between Native and non-Native languages, on the other. In her analysis, a more honest anthropology recognizes the productive and open-ended uncertainties that derive from misunderstandings that

are qualified as "communicative disjuncture" (27), "partial connections" (31), and "equivocation" (212) in the points of contact between
Native and non-Native ontologies. But what of the role of historically
unfolding relations of stratified, inequality-producing power that suf-
fuses and saturates those points of contact?

The distinct ontological world in which Native peoples in California
live and experience the distinctive roles of non-human sentient beings
is ensconced in, and some might say over-determined by, social, economic, and political structures repeatedly disrupted and reassembled
through cascading regimes of colonialism and capitalist transformation.
Much as I might, with the inspiration of Kohn's (2013) volume, aim
to comprehend the world as non-human sentient regalia do, it would
remain impossible to even initiate such a project except (trapped)
within the parameters of those transformations. As Bessire and Bond
note, Mario Blaser's work (see Blaser, 2013) constitutes an effort to
bring political histories to bear upon ontological anthropology, affirming the "co-construction of worlds through violent and conflictive histories, surging through sameness and difference alike" (Bessire & Bond,
2014: 451n10), all the while insisting that Indigenous histories, stories,
and relations with non-human beings "are not easily brought into the
fold of modern categories" (548).

Blaser's political ontology may not intersect with how I am inter-
preting the fate of pre-Columbian gold. However, like him, I would not
insist that gold is simply an objectively real entity about which there
are multiple cultural interpretations, identified in Blaser's approach
as a major failing of the culture concept in anthropology. While agree-
ing with Blaser and others that alterity means that modernity is not,
can never be, all encompassing, my sense of what "political" means
leads me back to what exactly happens to Indigenous practices and
relationships with gold, regalia, and other entities in colonial, and also
in pre-European, transformations. In the work of Nicholas J. Saunders
(1999; 2003), Carl Langebaek (1989; 1999; 2003), Warwick Bray (2003;
2005), Ana Maria Falchetti (2000; 2003) and others, Indigenous relationships with gold are focused upon: first, the transformative histories
of gold prior to European contact; and, second, the dynamic ways in
which Amerindian peoples attempted to understand and interact with
European ontologies after contact in ways that do not strictly conform
to either resistance or accommodation, and certainly not to a rigid,
impermeable boundary of ontological incommensurability.

I loosely group the literature I will review into two areas of analysis: one area that attends to the dynamic meaning of substances
among which gold played a role in pre-Columbian Colombia and a

second area focused upon the dynamically changing production of particular sorts of objects made of gold during the thousands of years of pre-Columbian metallurgy. In both cases, authors are concerned to emplace gold, both as a substance in itself and as particular objects, in larger, encompassing systems of producing meaning and materiality, which were not stable during the centuries before Europeans invaded WINC. Saunders (1999; 2003) draws attention to "the interplay between indigenous notions of power and the materiality of the pan-Amerindian 'aesthetic of brilliance' within which gold and its alloys were variably located" (Saunders, 2003: 16). He perceives that a central parameter of Amerindian perception before European conquest was "spiritual brilliance," elaborated into what he calls "philosophies of light":

> [T]he sacred brilliance of natural phenomena, ritual knowledge, glowing oratory, shiny matter, and technology were fused as one. Power, object, technological process, and enchantment were inextricably linked in the ebb and flow of cosmic forces in a universe conceived and governed by the symbolic propensities of analogical reasoning. (36)

Two essential elements for my argument follow from Saunders's discussion of the semiotic and political power of light as captured in and by shiny substances and the brilliant objects made from them. First, Saunders argues that the development of metallurgy and the use of gold and gold alloys in pre-Columbian history were preceded by hundreds, if not thousands, of years of use and deployment of other shiny substances:

> Indigenous Amerindians' valuations of gold, silver, and their alloys derived from already established ideas concerning the aesthetic of brilliance that hitherto had been connected to minerals, shells, plants, animals, and natural phenomena as they appear in nature and, transmuted through technology, as artifacts. From this perspective, metals were received into a pre-existing, age-old, symbolic, analogical, and multisensory world of phenomenological experience that has little in common with fifteenth-century European or modern notions of commercial wealth. (23)

That objects made of metals such as gold and gold alloys fit into pre-existing pre-Columbian ontologies, according to Saunders, does not deny that the technologies for purifying and casting metal objects were highly innovative, socio-politically transformational, and strongly supportive of the perceived power of shiny objects made from such new materials. Indeed, the manner in which whole societies and regions came to utilize gold and gold alloys to denote spiritual and political power instead of substances they had previously used in such a manner – as appears to be the case with green jade from lower Mesoamerica in what is now Costa Rica – underscores for Saunders the dynamic, unstable nature of pre-Columbian ontologies. "[T]he ideologies of gold," he writes, "emerged from prior ideologies associated with non-metallic brilliant matter" (29), in the end replacing the deployment specifically of jade. Such dynamism, in turn, conditioned a second outcome: when the Spanish arrived, their Indigenous interlocutors in Mesoamerica and the Andes accommodated the shiny objects the newcomers brought – glass and unfamiliar metals – within the existing system of sacred brilliance. In standard histories of gold and the dominant narratives about the Spanish *conquista*, this encounter is always told from the perspective of the wily European who successfully hoodwinks vast treasures from ignorant, brutish Natives with only a few relatively worthless trinkets. Saunders's analysis reasserts the agency of the Native interlocutors, who see the exchange of gold and gold alloy objects (as well as pearls, emeralds, jade, and other objects of brilliance) for glass beads as one of like for like and a means to bring what the Europeans had to offer into the existing dynamic system of spiritual and political power. Rather than incommensurable difference between Western and Amerindian ontologies, which Bessire and Bond imply inevitably locks Native ontology in a provincial or parochial closet, such a view reclaims Native ontologies as global ontologies that have also interacted with non-Native ontologies on their own terms.

Among two contemporary Indigenous groups in Colombia, the Kogi and the Arhuaco of the Sierra Nevada de Santa Marta massif, a significant number of pre-Columbian gold objects are still in the possession of ritual specialists (known as *mamas* among contemporary Kogi), who use them in ceremonies that maintain historical relationships with the pre-Columbian civilization in the region that has come to be known as Tairona. Gerardo Reichel-Dolmatoff (1985), the renowned don of Colombian anthropology and archaeology, was among the first to document the existence and use of pre-Columbian gold artefacts in Kogi rituals. Beginning in the 1950s, he collected and analysed Kogi lifeways in his two-volume opus, *Los Kogi: Una tribu de la Sierra Nevada de Santa Marta*. The Kogi, in particular, have received media attention because the Western gaze focused on ceremonial life (and the use of gold objects) intersects with common sense, essentialized perspectives on Native religion and "tradition" embodied in the widely distributed and viewed 1991 film *From the Heart of the World: The Elder Brothers Warning* (see Ereira 1992). Archaeologists Warwick Bray (2003) and Augusto

Oyuela-Caycedo (2002), among others, have substantiated the important relationships between Tairona and Kogi material culture, architecture, and symbolism. The work of Colombian cultural anthropologists,
including Juan Carlos Orrantia (2002) and Diana Bocarejo Suescún
(2002), critiques the popularized essentialist view of Kogi tradition as
an unchanging and stabilized Indigenous world-in-amber, while showing that Kogi and Arhuaco communities make sense of the contemporary world in which they live precisely because of the agility of their
Native ontologies. As Bocarejo Suescún writes, Kogi are "recreating
indigenous difference through contact" (2002: 4) with the non-Native
world on an ongoing basis. Contemporary Kogi ontologies, specifically
with respect to pre-Columbian gold artefacts, do not persist, therefore, because of parochial Kogi isolation from the non-Native world of
gold-as-money, gold-as-ornament, gold-as-love, but in intimate cogni-
zance of those non-Native deployments of gold – a proximate alterity
that is far from stable, temporally or semiotically.

Langebaek's work offers a different historicization of pre-Columbian
gold ontology, focused on dynamically changing production of particular sorts of objects made of gold during the hundreds of years of
pre-Columbian metallurgy. While Langebaek agrees with Saunders
and others that "[p]re-Columbian metallurgy clearly belongs in the domain of religious ideology" (Langebaek, 2003: 247), and similarly for
contemporary peoples such as the Kogi as well, his analysis notes that,
during different time periods, political, social, and economic systems
diverged in WINC, conditioning the production of distinctly different
sorts of gold objects. From roughly 500 BC to AD 900, he points out:

> Early developments in goldwork and the elaboration of unique (and fre-
> quently impressive) adornments are consistent with a social context in
> which power was a highly personalized and institutionalized affair and
> in which leaders completely lacked the mechanisms to make status inher-
> ited. (248–9)

During the early period, the production under elite control of large,
sophisticated, and unique gold objects, such as statuary, musical instruments, and large *poporos* (containers in which the lime used in chewing
coca is kept) was a means by which those elites laid claim to power
and negotiated their deployment of it in competition with other elites.
Control of such objects did not signify the accumulation of personal
wealth but rather of spiritual qualities that corresponded to political
power on display. Graeber writes: "[O]bjects whose value is seen to lie
in their particular histories or identities have an equally strong tendency

to be assimilated to the social identity or persona of their owners, thus generating the impulse to show them off" (2001: 104). Chemical analysis reveals that many of these objects were fashioned out of very high gold content alloys. Langebaek notes a marked transition after AD 900 away from the production of remarkable, unique gold objects and towards mass production of standardized personal ornaments, such as earrings, nose rings, and breastplates, in specialized artisanal workshops amidst an apparently growing population that had access to such gold objects. In the central highlands of WINC, the Muisca civilization specialized in the mass production of small votive figurines, known as *tunjos*, which were ritually emplaced in caves, agricultural fields, homes, near wells, or in groves of trees by farmers and other non-elites. Mass-produced objects were often made of the gold-copper alloy tumbaga; Falchetti (2003) argues that, in WINC, alloys were not considered less beautiful or important than pure gold, since their coppery shimmer was also attributed with tremendous aesthetic value, both visual and olfactory. Perhaps, however, tumbaga as a raw material was more available to metallurgical craftsmen. Indeed, lest European sensibilities conclude that these changes signified a decline in craftsmanship, skill, or the esteem that gold and gold alloys possessed in pre-Columbian Colombia, Langebaek states that, whereas in previous archaeological analysis "modifications in goldwork [were] interpreted as indicative of late, less-developed groups ... [they] are here interpreted as indicating social transformations from ideological means of control to more institutionalized political organizations in which power relied on the control of labor and resources" (2003: 249).

The implications of Langebaek's historicization of the development of goldwork production, like Saunders's historicization of gold's semiotics, underscore the unstable and unbounded character of pre-Columbian ontologies and also their contemporary descendant ontologies among peoples such as the Kogi. I conclude that such archaeological and ethnological analysis highlights the manner in which ontological difference is propelled by the interaction between dissimilar ontologies rather than by their parochialization or warehousing in separate and sharply bordered conceptual territories, which Blaser (2013) also disavows. If the mainstream histories of gold with which I began this discussion simply elide the radically different ontologies of gold in WINC and those that existed elsewhere in this hemisphere before the European invasions, ontological anthropology may do no better justice to the complex interactions between Amerindian and European ontologies in which a substance, gold, has never been a single "it." In the next section, I probe deeper: if gold is not a single "it" in any single ontology much less in the interaction between ontologies, how might gold, as a taken-for-granted actor in mainstream histories, be additionally reconceptualized from different theoretical perspectives?

Gold the Actor

Part of gold's common-sense actor identity in mainstream histories derives from its allegedly sui generis characteristics, with respect to its alleged indestructability and immortality. One elite jewellery website confides:

> Pure gold is indestructible, it will not corrode, rust or tarnish and cannot be destroyed by fire. All the gold taken from the earth during all of recorded history is still being melted and re-melted and used again and again. The gold you own and wear today may once have adorned King Solomon's Temple, perhaps been worn by Cleopatra or have been carried to Bethlehem by three wise men. (Gordon, n.d.)

All the gold that has ever been mined, we are told, is still bouncing around today, melted and reshaped, stowed away in treasuries or parading about in wedding rings. Unfortunately, the same is exactly true about mercury – all the mercury that has ever been mined or extracted is still with us today. It cannot be destroyed. Unlike gold, mercury can and does easily form a variety of compounds and, in that way, moves from solid to liquid to gaseous states, but it is still around, still poisoning biological systems everywhere it infiltrates, still becoming ever more concentrated as animals are nourished up the food chain. Problematically and ironically, the extraction of elemental gold from mineral ore has been facilitated using mercury for possibly up to three thousand years. Thanks to human relations with gold, more mercury is in the biosphere than there ever would have been without these arrangements. (For a succinct instantiation of the mercury-gold dyad in the historic Sierra Nevada gold-mining region of California, see Alpers, Hunerlach, May, & Hothem, 2005.) If such facts weren't enough to tarnish the romance of "indestructability and immortality," plutonium's most stable isotope has a half-life of eighty million years. That period of time is not equivalent to the indestructability evoked by jewellers, but it is close enough. The list of virtually "immortal" toxic substances – dioxin, polystyrene – is far too long.

Ignoring gold's elemental persistence, the fact that it does not corrode or decay, as well as its other unique characteristics that are said to explain its historical actor role, may not be feasible, or even useful, but such characteristics need not be viewed as inherently or commonsensically positive, remarkable, or agentive. Such a perspective, in the view of Marx, fetishizes an object, such that it seems to be an autonomous actor, as Graeber explains:

> In fetishizing an object, then, one is mistaking the power of a history internalized in one's own desires, for a power intrinsic to the object itself. Fetish objects become mirrors of the beholder's own manipulated intentions. (Graeber, 2001: 115)

Jane Bennett (2010), a premier theorist of "vibrant matter,"[3] has critiqued the Marxist perspective regarding the fetishism of objects because of the limits to the efficacy of demystification, "that most popular of practices in critical theory, [which] should be used with caution and sparingly, because demystification presumes that at the heart of any event or process lies a human agency that has illicitly been projected into things" (xiv). In her effort to construct an ontology that recognizes and explores agency in the non-human and in objects, Bennett argues:

> What demystification uncovers is always something human, for example the hidden quest for domination on the part of some humans over others ... Demystification tends to screen from view the vitality of matter and reduce *political* agency to *human* agency. Those are tendencies I resist. (xv; emphasis in the original)

It is important to note here that gold's vibrancy and agentive character could not, from the perspective of Bennett's theorization of vibrant matter, be in any way unique or indeed special. Lead is just as vibrant as gold – and, more apropos, so is mercury, whose agency as a toxic substance that persists, accumulates, crosses boundaries, and insidiously poisons comprises a vibrancy quite as vivid as gold. Thus, it is not the agency of metals that I contest in Bennett's work, but rather the diminished utility of a Marxist deployment of demystification for understanding the agency of a non-human entity, in this case, gold. I would argue that a Marxist analysis rooted in Bertell Ollman's (1978) exegesis of the concept of alienation does not incur such a restricted form of demystification. My reading of Ollman offers a richly complex understanding of fetishism, which does not preclude a consideration of the vibrant nature of matter, such as gold, that Bennett has developed. By the same token, as I explain further on, Bennett does not entirely exclude the manner in which human labour shapes the agency of non-human matter or objects. More to the point, Ollman's elucidation of Marx's position does resonate – in part – with Bennett's characterization:

> The "fetishism of commodities" refers to people's misconception of the products of labor once they enter exchange, a misconception which accords these forms of value leading roles in what is still a human drama ... Misreading this story as one about the activities of inanimate objects, attributing to them qualities which only human beings could possess, positing living relations for what is dead, is what Marx call "the fetishism of commodities." (Ollman, 1978: 195)

If gold objects, not to mention gold itself, are "dead" and their value in whatever ontology only possible through reified misperception, how is one to reconcile with Bennett's "vibrant" matter? Marx's understanding of the fetishism of commodities was in many ways primarily directed towards the manner in which money, and money being deployed as capital, obscured the origin of value in human labour and therefore mystified the production of surplus value, which is, for Marxists, at the crux of the capitalist economic system. Since gold, in Western ontologies, is inseparable from its historical relationship with money, gold presents a particularly obvious target for demystification. Yet Marx, as Ollman explains, was highly aware of the decisively real (that is, reified) force that money, the ultimate fetishized commodity, exercised over human beings. Marx wrote that money is "changed into a true god, for the intermediary reigns in real power over the thing that it mediates for me. Its cult becomes an end in itself" (quoted in Ollman, 1978: 200). In this light, Ollman argues:

> People acquire their conception of reality from what they experience ... and their conception of reality helps determine what they experience ... The power of capital, or of any of the workers' products, over the worker always reflects the power of the people who dominate it and use it as an instrument. However, *through reification and inside the context of capitalism, capital may itself exercise certain powers*. Marx is not guilty of the fetishism he discovers in capitalist societies, *because the powers he ascribes to products are never considered theirs as natural qualities*. (200; emphasis added)

If these insights are understood in light of Raymond Williams's (1977) qualification of ideas, aesthetics, philosophies, ideologies, and daily life itself as products, no less manufactured under specific conditions with specific relations of production using specific forces of production than cars, computers, or chairs, it is possible to understand both gold the material and objects made of gold as actors; this understanding, of course, is true for gold-as-money as well. Graeber offers cautionary notes to such a move, with respect to both gold-as-money and unique gold objects, which he discusses under the category of "heirlooms," a category that evokes the early period of gold metallurgy in pre-Columbian history:

> The value of an heirloom is really that of actions: actions whose significance has been, as it were, absorbed into the object's current identity ... Since the value of the actions has already been fixed in the physical being of the object, it is perhaps a short leap to begin attributing the agency behind such actions to the object as well. (2001: 105)

When gold is used as money, by contrast, the ontological framework shifts drastically. Coins used as currency, rather than as ornament, should never consist of unique or singular specimens. Coins when used only as currency should be completely generic; they should "present a frictionless surface to history" (94). Gold acting as money, as Marx argued, realizes its specificity only through acts of exchange that are in the future rather than in the past. Yet, following Marx, when gold manifests as either unique objects, or "heirlooms," or when gold is used as money, its activities are experienced by human beings as if gold in these forms were an actor. In experiencing gold in that way, gold's agency is powerfully real, not misconstrued. It is thus possible from a Marxist perspective to concede the agency of gold, as long as its agency is not understood as natural or pre-ordained.[4]

Bennett also writes about metals – as substances in and of themselves – as actors from a human-centred perspective. In her critique of the "hylomorphic" model of material vitality (about which I am not in the least competent to comment upon), she notes:

> The hylomorphic model is ignorant of what woodworkers and metallurgists know quite well: there exist "variable intensive affects" and "incipient qualities" of matter that external forms [can only] bring out and facilitate ... Artisans (and mechanics, cooks, builders, cleaners, and anyone else intimate with things) encounter a creative materiality depending on the other forces, affects or bodies with which they come into close contact. (2010: 56)

Though she is at pains to distance this observation from the "historicity of objects" and the "social lives of objects,"[5] subjects relegated to "anthropology, sociology and social sciences" (57), Bennett returns to the observation shortly thereafter to note:

> But metal is always metallurgical, always an alloy of the endeavors of many bodies, always something worked on by geological, biological, and often human agencies. And human metalworkers are themselves emergent affects of the vital materiality they work ... The desire of the craftsperson to see what a metal can do, rather than the desire of the scientist to know what a metal is, enabled the former to discern a life in metal and thus, eventually, to collaborate more productively with it. (60)

Attention to the activities of artisans and the unfolding of craft necessitates an equivalent focus upon the historicity and social life of things, I would argue. But taking Bennett at face value here indicates that the analysis of "vibrant matter" does not preclude the heavily reified interaction between substance/material/object and humans as at least one condition for the former's very real agency. Anna Tsing's (2015) biography of the matsutake mushroom, as heroic a non-human entity as imaginable, vigorously maintains the multiply complex character of this bit of vibrant matter, vividly portraying the mushroom as an evolutionary and ecological agent in its own right and as a cog in an international commodity chain, always shaped by human relationships. Annemarie Mol, in her ethnography of atherosclerosis in a Dutch hospital, also explores the agency of objects and writes that "no object, no body, no disease is singular" (2002: 6). She asserts that "objects, in their turn, are not taken here as entities waiting out there to be represented but neither are they the constructions shaped by the subject-knowers" (32). She goes on to dispute the dichotomies between "subject-humans" and "objects-nature" on the one hand, and between "actively knowing-subjects" and "objects-that-are-known" on the other, arriving at two very odd and provocative formulations, which I would argue resonate with Ollman's discussion of reified objects that can manifest as actors: first that "like (human) subjects, (natural) objects are framed as parts of events that occur and plays that are staged. If an object is real this is because it is part of a practice" (Mol, 2002: 44); and, second, that "a way out of the dichotomy between knowing subject and the objects-that-are-known [is] to spread the activity of knowing widely ... Instead of talking about subjects knowing objects we may then, as a next step, come to talk about enacting reality in practice." (50)

This negotiation is not for Mol, nor for my discussion, a sort of parlour game that animates objects for the pleasure of those who can literally afford to see the world in such a way. For Mol, the stakes are potently significant understandings of disease and treatment. For gold, as for other metals, minerals, and extracted natural resources, the stakes are an attempt to understand a world coming into being that may no longer be hospitable for life. These stakes have changed dramatically – in gold's case from a long early period in gold's relationship with humans during which pure metal was accessed on the earth's surface or in rivers beds to increasingly, and currently exponentially, more damaging and destructive forms of extraction, the effects of which are almost completely hidden from those consuming gold, at least until the toxic effects are simply unavoidable. In Michelle Murphy's work regarding buildings and illness, she focuses upon both the perceptibility and imperceptibility of objects. She writes: "The history of how things come to exist is intrinsically linked to the history of how things come not to exist" (2006: 9). Bessire and Bond have stated, with respect to petroleum, in particular, and as analysed in Bond's ethnographic work (Bond, 2013) but in a manner broadly applicable to all extracted resources:

> The materiality of hydrocarbons also demands a more careful analysis than Marxism allows. In mutated ecologies, cancerous bodies, scarred landscapes, and contorted weather patterns, *the force of hydrocarbons surpasses the labored dimensions of a commodity* ... When consumed, hydrocarbons do not disappear but come to structure apocalyptic forms of obligation that may exceed the capacities of life itself. (Bessire & Bond, 2014: 451; emphasis added)

Recent anthropological work in and around the notorious Ok Tedi mine in Papua New Guinea (see Golub, 2014; Kirsch, 2014) epitomizes the attack on life itself that contemporary gold-mining and purification have become.

Ultimately, I contend that this conclusion is the meaning and reason for analyses of gold that simultaneously deconstruct gold as the protagonist of mainstream histories and reassert gold's (aggressively) agentive identity as a multiplicity not completely under the control of human subjectivity, intentionality, or foresight. Taussig's (2004) disturbing protest, at times a tirade, against Bogota's Museo del Oro (the Gold Museum) echoes the perils of that multiplicity. The Museo del Oro *is* indeed a deeply ambivalent institution (see Gaitán, 2006; Field, 2012), which appears to embrace Colombia's Indigenous history, but in doing so erases contemporary Indigenous peoples. More to the point, it deploys gold objects as archaeological heritage when, in fact, all of the items on display were unearthed by *guaquería*, or illicit, non-scientific excavation. Taussig's (2004) critique hinges on what the Museo is not in the slightest bit informative about – the post-conquest relationship between gold and gold-mining, on the one hand, and the legacies of slavery and the racism of contemporary Colombia, on the other. In view of those inextricable relationships forged since the 1500s, Taussig considers gold a "transgressive substance," and he weaves together an acerbic journey through gold's relationships with slavery and suffering, with the history of cocaine in the late twentieth and early twenty-first centuries. Taussig identifies both gold and cocaine not so much as "vibrant matter" as toxic entities, fetishes to be sure, affirming their agency, even their potential personhoods. His book offers a hall of distorted, perverted mirrors in which the many forms of gold – the many "its" of gold – are neither part of a larger whole nor separable from one another.

Conclusion

Gold is not a singular it-as-actor. In many (not all) societies during the last 6,500 or so years, there is a "there," a sociocultural place so to speak, where gold exists and transformatively changes over time. The "there" exists. But there is no "it."

The dominant histories of gold construct a tidy argument about gold's role in history that requires a silence around the other ontological worlds where gold has been used and recognized, not to mention the human histories during which gold had very little ontological significance. My discussion of incommensurable ontologies argues that gold is a disjointed and dynamic roster of entities and identities defined only by their relationship to humans rather than by gold's putative immanent qualities or characteristics. In WINC, gold was both an element in a dynamic ontology of the shiny and a raw material whose sociocultural utility and significance transformed substantially over a series of hundreds of years.

The discussion of vibrant materialities is exciting but must, I would argue, abjure the alienation of human agency, even if writers seek to pay less (or very little) attention to human agency than, for example, Marxist approaches. Using Raymond Williams's (1977) comprehensive description of production, one must refuse to separate gold's materiality from its ideation; however, a lens focused upon the production of gold – as pure substance, as objects, and as meanings – does not completely account for gold's entire agency. Therefore, the agency of this material must be considered in new ways that Marxism does not provide. However, gold's agentive quality must be divorced from assumed or a priori immanent qualities that the dominant histories always draw upon and instead derive, at least in part, from a focus upon the unforeseen, physiochemical materiality of mining, extraction, and, ultimately, environmental destruction.

Notwithstanding the powerful domination of the "history of gold" and the recurring waves of gold fanaticism in financial markets, in the world of jewellery, and around the symbolism of love and commitment, gold's fictive "it-ness" is not, I hope, a permanent fixture into an endless future. Veronica Davidov's (2013) complicated account of the complex rejection of, revulsion against, and simultaneous economic deployment of gold in the Soviet Union underscores that humans can and will continue to imagine extinguishing both gold's "it" and its "there." The agency of gold will not change, much less vacate, its position in relation to humans without mighty feats of imagination.

> When we win on the global scale, I think, we will build public toilets out of gold in several largest cities in the world. (V.I. Lenin, quoted in Davidov 2013: 1)

Acknowledgments

The genesis of this chapter traces back to a semester I spent at and sponsored by the Universidad de los Andes in 2009, funded by a Fulbright Scholar Program Fellowship. I would like to profusely thank Universidad de los Andes and the generosity and brilliant work of my sponsor there, Carl Henrik Langebaek Rueda. Many thanks to the curator-scholars in the Museo del Oro in Bogota, especially Clara Isabel Botero, as well as the curator-scholars in the branches of the Museo del Oro in Cali and Santa Marta. My colleagues and friends Cristobal Gnecco and Joe Watkins have been, as always, sources of analysis and insight. It has been a great pleasure to work with my colleagues and friends Elizabeth Ferry and Andrew Walsh on the multitudinous facets of the anthropology of precious minerals, and I look forward to the next phase of our collaboration. Lastly, I thank Vernon and Berta Welch of the Wampanoag Tribe of Gay Head (Aquinnah) for introducing me to the world of wampum and the Muwekma Ohlone Tribe, especially Rosemary Cambra and Alan Leventhal, for continuing inspiration.

NOTES

1 I use the term "actor" in the way that Anthony Giddens deployed the term in Giddens, 1983."Actant," used in Latourian analysis, is another possible option. I understand from colleagues who write within actor-network theory (ANT) frameworks that Latour and those with whom he works reject eclectic theorizing. Even though I am doing quite a bit of that in this chapter, I will respect the terms of ANT theorists, at least to the extent of not using "actant," since I am not consenting to limit my theorizing to only ANT frameworks.

2 The oldest gold objects yet unearthed come from the Varna Necropolis in what is now Bulgaria and are dated to about 4,500 BC (Bachman, 2006).

3 Jane Bennett and Tim Ingold, two of the most prominent theorists in the field of "vibrant matter" (Bennett, 2010), compellingly describe how seeing materiality as "vital" transforms our understanding, ethics, and politics concerning humans and non-humans. Things exert their pressure in creating worlds; as Mario Blaser puts it, "reality is always in the making through the dynamic relations of heterogeneous assemblages involving more-than-humans," a process he calls "worlding" (Blaser, 2013: 54). Both Bennett and Ingold draw from Deleuze and Guattari's understanding of, in Ingold's paraphrasing, "the way in which materials of all sorts, energized by cosmic forces and with variable properties, mix and meld with one another in the generation of things" (Ingold, 2010: 92; Deleuze & Guattari, 1987).

4 See Edwards (2010) for an affirmation of the utility of Marxist analysis in light of "new materialist" approaches.

5 My own previous work regarding gold certainly falls under the genres of the "historicity of objects" and the "social lives of objects" (see Field 2008; 2012; 2016).

REFERENCES

Alberti, B., Fowles, S., Holbraad, M., Marshall, Y., & Whitmore, C. (2011). "Worlds otherwise": Archaeology, anthropology, and ontological difference. *Current Anthropology*, *52*(6), 896–912. https://doi.org/10.1086/662027

Alpers, C., Hunerlach, M., May, J., & Hothem, R. (2005). *Mercury contamination from historic gold mining in California*. United States Geological Survey (USGS) Fact Sheet 2005–3014. Dixon, CA: USGS Publications. Retrieved from https://pubs.usgs.gov/fs/2005/3014/fs2005_3014_v1.1.pdf

Bachmann, H. (2006). *The lure of gold: An artistic and cultural history*. New York: Abbeville Press.

Bennett, J. (2010). *Vibrant matter: A political ecology of things*. Durham, NC: Duke University Press.

Bernstein, P.L. (2000). *The power of gold: The history of an obsession*. New York: John Wiley & Sons.

Bessire, L., & Bond, D. (2014).Ontological anthropology and the deferral of critique. *American Ethnologist, 41*(3), 440–56. https://doi.org/10.1111/amet.12083

Blaser, M. (2013). Ontological conflicts and the stories of peoples in spite of Europe: Toward a conversation on political ontology. *Current Anthropology, 54*(5), 547–68. https://doi.org/10.1086/672270

Bocarejo Suescún, D. (2002). Indigenizando "lo blanco": Conversaciones con arhuacos y koguis de la Sierra Nevada de Santa Marta. *Revista Antropología y Arqueología, 13*, 3–44.

Bond, D. (2013). Governing disaster: The political life of the environment during the BP oil spill. *Cultural Anthropology, 28*(4), 694–715. https://doi.org/10.1111/cuan.12033

Bray, W. (2003). Gold, stone, and ideology: Symbols of power in the Tairona tradition of Northern Colombia. In J. Quilter & J.W. Hoopes (Eds.), *Gold and power in ancient Costa Rica, Panama, and Colombia: A symposium at Dumbarton Oaks, 9 and 10 October 1999* (pp. 301–44). Washington, DC: Dumbarton Oaks.

Bray, W. (2005). Craftsmen and farmers: The archaeology of the Yotoco Period. In M. Schrimpff (Ed.), *Calima and Malagana: Art and archaeology in Southwestern Colombia* (pp. 98–139). Bogota: Pro Calima Foundation.

Davidov, V. (2013). Soviet gold as sign and value: Anthropological musings on literary texts as cultural artifacts. *Etnofoor, Gold, 25(1)*, 15–28.

De La Cadena, M. (2015). *Earth beings: Ecologies of practice across Andean worlds*. Durham, NC: Duke University Press.

Deleuze, G., & Guattari, F. (1983). *Anti-Oedipus: Capitalism and schizophrenia*. Minneapolis: University of Minnesota Press.

Deleuze, G., & Guattari, F. (1987). *A thousand plateaus: Capitalism and schizophrenia*. Minneapolis: University of Minnesota Press.

Edwards, J. (2010). The materialism of historical materialism. In D. Coole & S. Frost (Eds.), *New materialisms: Ontology, agency, and politics* (pp. 281–98). Durham, NC: Duke University Press.

Ereira, A. (1992). *The elder brothers: A lost South American people and their message about the fate of the earth*. New York: Knopf.

Falchetti, A.M. (2000). The gold of greater Zenú: Prehispanic metallurgy in the Caribbean lowlands of Colombia. In C. McEwan (Ed.), *Precolumbian gold: Technology, style, and iconography* (pp. 132–51). London: British Museum Press.

Falchetti, A.M. (2003). The seed of life: The symbolic power of gold-copper alloys and metallurgical transformations. In J. Quilter & J.W. Hoopes (Eds.),

Gold and power in ancient Costa Rica, Panama, and Colombia: A symposium at Dumbarton Oaks, 9 and 10 October 1999 (pp. 345–81). Washington, DC: Dumbarton Oaks.

Field, L.W. (2008). *Abalone tales: Collaborative explorations of sovereignty and identity in Native California*. Durham, NC: Duke University Press.

Field, L.W. (2012). El sistema del oro: Exploraciones sobre el destino (emergente) de los objetos de oro precolombino en Colombia. *Antipoda: Revista de Antropología y Arqueología, 14*, 67–94. https://doi.org/10.7440/antipoda14.2012.04

Field, L.W. (2016). Dynamism not dualism: Money and commodity, archaeology and guaquería, gold and wampum. In L. Field, C. Gnecco, & J. Watkins (Eds.), *Challenging the dichotomy: The licit and the illicit in archaeological and heritage discourses* (pp. 180–96). Tucson: University of Arizona Press.

Gaitán, A.F. (2006). Golden alienation: The uneasy fortune of the Gold Museum in Bogotá. *Journal of Social Archaeology, 6*(2), 227–54. https://doi.org/10.1177/1469605306064242

Giddens, A. (1983). *Profiles and critiques in social theory*. Berkeley: University of California Press.

Golub, A. (2014). *Leviathans at the gold mine: Creating indigenous and corporate actors in Papua New Guinea*. Durham, NC: Duke University Press.

Gordon, K. (n.d.). Understanding gold. *Kenneth Gordon, Private Jeweller* (website). Retrieved from http://www.kennethgordon.net/knowledge/gold/

Graeber, D. (2001). *Toward an anthropological theory of value: The false coin of our own dreams*. New York: Palgrave.

Green, T. (1993). *The world of gold: The inside story of who mines, who markets, who buys gold*. London: Rosendale Press.

Hart, M.H. (2013). *Gold: The race for the world's most seductive metal*. New York: Simon& Shuster.

Holbraad, M. (2012). *Truth in motion: The recursive anthropology of Cuban divination*. Chicago: University of Chicago Press.

Ingold, T. (2010). The textility of making. *Cambridge Journal of Economics 34*(1): 91–102. https://doi.org/10.1093/cje/bep042

Kirsch. S. (2014). *Mining capitalism: The relationship between corporations and their critics*. Berkeley: University of California Press.

Kohn, E. (2013). *How forests think: Toward an anthropology beyond the human*. Berkeley: University of California Press.

Kwarteng, K. (2014). *War and gold: A 500-year history of empires, adventures and debt*. New York: Public Affairs.

Lame Deer, J., & Erdoes, R. (1972). *Lame Deer, seeker of visions*. New York: Washington Square Press.

Langebaek, C.H. (1989). El uso de adornos de metal y la existencia de sociedades complejas: Una visión desde Centro y Suramérica. *Revista de Antropología y Arqueología, 71*, 73–90.

Langebaek, C.H. (1999). Pre-Columbian metallurgy and social change: Two case studies from Colombia. In G.G. Politis & B. Alberti (Eds.), *Archaeology in Latin America* (pp. 244–70). London: Routledge.

Langebaek, C.H. (2003). The political economy of pre-Columbian goldwork: Four examples from Northern South America. In J. Quilter & J.W. Hoopes (Eds.), *Gold and power in ancient Costa Rica, Panama, and Colombia: A symposium at Dumbarton Oaks, 9 and 10 October 1999* (pp. 245–78). Washington, DC: Dumbarton Oaks.

Mol, A. (2002). *The body multiple: Ontology in medical practice.* Durham, NC: Duke University Press.

Murphy, M. (2006). *Sick building syndrome and the problem of uncertainty: Environmental politics, technoscience, and women workers.* Durham, NC: Duke University Press.

Ollman, B. (1978). *Alienation: Marx's conception of man in capitalist society.* Cambridge: Cambridge University Press.

Orrantia, J.C. (2002). Esencialismo desde el corazón del mundo: Información como legitimación del riesgo. *Revista Antropología y Arqueología, 13*, 45–77.

Oyuela-Caycedo, A. (2002). El surgimiento de la rutinización religiosa: La conformación de la elite sacerdotal Tairona-Kogi. *Revista de Arqueología del Área Intermedia, 4a*, 45–64.

Reichel-Dolmatoff, G. (1985). *Los Kogi: Una tribu de la Sierra Nevada de Santa Marta, Colombia.* Bogotá: Procultura, Nueva Biblioteca Colombiana de Cultura.

Saunders, N. (1999). Biographies of brilliance: Pearls, transformations of matter and being, ca. AD 1492. *World Archaeology, 31*(2), 243–57. https://doi.org/10.1080/00438243.1999.9980444

Saunders, N. (2003). "Catching the light": Technologies of power and enchantment in pre-Columbian goldworking. In J. Quilter & J.W. Hoopes (Eds.), *Gold and power in ancient Costa Rica, Panama, and Colombia: A symposium at Dumbarton Oaks, 9 and 10 October 1999* (pp. 15–47). Washington, DC: Dumbarton Oaks.

Skafish, P. (2013). From anthropology to philosophy: Introduction to Eduardo Viveiros de Castro. *Radical Philosophy, 182*, 15–16.

Sutherland, C.H.V. (1959). *Gold: Its beauty, power, and allure.* New York: McGraw Hill.

Taussig, M. (2004). *My cocaine museum.* Chicago: University of Chicago Press.

Tsing, A.L. (2015). *The mushroom at the end of the world: On the possibility of life in capitalist ruins.* Princeton, NJ: Princeton University Press.

Vilar, P. (1976). *A history of gold and money, 1450–1920*. London: New Left Books.

Viveiros de Castro, E. (1998). Cosmological deixis and Amerindian perspectivalism. *Journal of the Royal Anthropological Institute, 4*(3), 469–88. https://doi.org/10.2307/3034157

Viveiros de Castro, E. (2013). Cannibal metaphysics: Amerindian perspectivism. *Radical Philosophy, 182*, 17–28.

Williams, R. (1977). *Marxism and literature*. Oxford: Oxford University Press.

Afterword: Facets of Preciousness

ANDREW WALSH, ELIZABETH FERRY,
AND ANNABEL VALLARD

The "preciousness" of precious minerals is, in a word, multifaceted. Obvious as it is, given the focus of this volume, the metaphor works. Considered from one angle, the quality of preciousness emerges in moments of engagement involving particular people and particular minerals: in a French museum visitor's appreciation of a one-of-a-kind specimen like Laurent (as discussed in Raveneau's chapter three), for example, or a Thai bead collector's contemplation of a new find (as discussed by Vallard in chapter four). A twist of the wrist offers another perspective, illuminating complex cultural and semiotic systems and regimes of value through which a "precious" gemstone might be variously characterized as "a girl's best friend" (as in a well-known song cited in the introduction), as having "an extraordinary charm that cannot be explained by words" (as in a Christie's catalogue description cited in Calvão's chapter five), or as a "BGY sapphire" with distinctive "spectroscopic properties" (as in a gemological report cited by Walsh in chapter two). Another twist exposes preciousness as a function of (or at least closely related to) the market value of particular minerals, and yet another as marking the primordial, something better felt than explained, still less as something to be bought or sold. And so on. In these concluding reflections, we advocate a perspective expansive enough to include these sometimes contiguous, sometimes seemingly disconnected facets of preciousness simultaneously.

Just as the facets of finished gemstones are products of lapidaries' work *with* minerals, the facets of preciousness we highlight here owe as much to the content of the preceding chapters as to our own initial approaches to them. As co-editors, we have been provoked by the cases collected here to think with fellow contributors about the matter of preciousness and the preciousness of at least some matter. While form demands that our summation of such provocations put a sequence to the different directions they have taken us, we do not mean to suggest priority; we intend the different views proposed here to be taken as alternative angles on the same thing. To begin, we return to what so engaged us in the scene with which we began the introduction – a roomful of anthropologists and mineralogists poring over specimens, loupes in hand – and to the careful contemplation of the inherently distinctive stuff of minerals themselves.

One of the problems with attempting to classify certain minerals or mineral specimens as "precious" is that the matter at hand so often expresses and/or is invested with distinctive qualities in ways that render it seemingly unclassifiable as anything but itself. As noted by Field in chapter six, for example, gold is commonly presented as different from any other substance, incomparable and unquestionably deserving of a starring role in its own history as well as in the history of humankind. Similarly, although the various mineral specimens, gemstones, and beads scattered through chapters by Walsh, Raveneau, Vallard, and Calvão are all subject to various classificatory schemes, their preciousness is unquestionably linked to their being seen as distinctive, singular, each one-of-a-kind in their own way. Citing Lucien Karpik (2010), Calvão in chapter five describes diamonds as "singular, incommensurable products" that are especially hard to evaluate, and none more so than those rarest of the rare "internally flawless" specimens of unverifiable origin. The singularity of sapphires, described by Walsh in chapter two, is of a different sort, evident in the interplay of colour and inclusions that have the potential to make each specimen memorable, affectively forceful, and generative of unique possibilities for evaluation and bargaining.

Singularity is perhaps most obviously a feature of minerals that have taken on social lives, identities, and even names of their own. For example, the Emerald Buddha, the Golden Jubilee Diamond, and the collected beads described in Vallard's chapter four and the personally named Alpine crystals described by Raveneau in chapter three reveal how social relations can stick to particular minerals in distinctive ways. As Raveneau puts it, their "singular character derives not only from the fact that [they are] particularly precious or unique, but also from the circumstances, contexts, and personal histories associated with [them]." In all cases, however, the processes involved in realizing this apparent singularity are anything but idiosyncratic – the skills of lapidaries who transform the unremarkable into something spectacular are as systematic, learned, and shared as are those of the mountaineers who gingerly remove crystals from Alpine "ovens" or those of gemologists whose expertise is invoked when evaluating particular specimens. And so, what is posited as inherently singular is revealed, through deeper consideration, as something more. As Field reminds us in chapter six, even gold – that most distinctive of substances – is "not a singular it-as-actor," but "a disjointed and dynamic roster of entities and identities defined only by their relationship to humans rather than by gold's putative immanent qualities or characteristics."

Writing in the early twentieth century, famed gemologist George Frederick Kunz described durability as another feature of "precious stones" that helps account for their preciousness, allowing them to manifest "permanence" in "a world of change" (1971: v). While such durability clearly matters in the valuation of certain minerals as precious, not least in the marketing of gemstones given as gifts, we are also reminded of Ponge's observation, noted in the introduction, that minerals are "the only thing[s] in nature that constantly [die]" (1972: 73). They do, in fact, change over time as, of course, do the engagements through which they come to be figured as "precious." Although commonly valued for their capacity to index timelessness, even the most precious of precious minerals can never be rendered fully finished, no matter the amount of processing and polish. The semiotic systems within which minerals make sense as "precious" in particular ways are themselves perpetually in formation, meaning that, as Field in chapter six notes is the case with gold in what is now Colombia, the "sociocultural utility and significance" of precious minerals can change substantially over time, whether gradually alongside the sociopolitical developments of Indigenous polities (as in the case Field describes) or relatively quickly (as in the case of conflict diamonds discussed by Calvão in chapter five). Just think of where the beads collected from burial sites in Thailand, as described in chapter four by Vallard, have been, where they are now, and where they might end up. Now imagine how they have been differently precious at various points in these long social lives. Or imagine, as Walsh suggests in chapter two, what observers of the far distant future might make of our times given the uneven, but clearly patterned, global distribution of sapphires sourced in northern Madagascar. Indeed, it is the durability of minerals that guarantees them futures of perpetual change and enduring uncertainty. We can see this future especially well in Bell's chapter one on scrapper videos, in which he notes that Ingold's "meshwork" approach (Ingold, 2012) and the framing of commodities as "global assemblages" by Stephen J. Collier and Aihwa Ong (2005) and Anna Tsing (2015) pertain to the dismantling of objects as well as to their composition. Such a focus on decomposition is particularly salient at a moment when recycling, upcycling, and other revaluations of waste are seen as increasing pertinent for scholars and an increasingly important livelihood for those outside of academia (Reno, 2016).

Durability is not a unique feature of minerals, of course. It is also a feature of the mountainsides, landscapes, ores, matrices, and other geological contexts within which museum quality specimens, gemstones, and gold can be found, not to mention a wide range of discarded electronics and toxic substances (such as mercury) that, as Field notes in chapter six, manifest "indestructability and immortality" in ways more clearly problematic than precious. It is only in combination with other mineralogical features, then, that durability figures in the valuation of preciousness. Consider, for example, the capacity of all the minerals considered in the preceding chapters to, as Walsh puts it in chapter two, "concentrate the economic value attributed to them in a remarkably versatile form." The small size and durability of sapphires mined in Madagascar, for example, means that miners can hide (or be imagined to hide) especially valuable specimens found in collectively mined pits in their mouths and that local traders, using strategies not too different from those of the diamond cartel De Beers, can amass and keep stores of them in anticipation of future sales and market shifts. These same qualities are also key to the global circulation of sapphires and other coloured stones, allowing, as Walsh puts it, the "transnational, and potentially untraceable, movements of great wealth" (see also Naylor, 2010; Brazeal, 2014; 2017).

As discussed in the introduction, brilliance, colour, lustre, and other features inspire associations between precious minerals and transcendent values. Many of these qualities only surface, however, in acts of skilled refining, processing, and representation through the specialized work of lapidarists (chapter four), gemstone labs (chapter two), and auctioneers (chapter five) that serves ultimately to make the preciousness of precious minerals apparent to those inclined to reckon it. At earlier stages in the commodity chain, however, what sets some minerals apart from others is their relative *in*accessibility – that is, their capacity to attract people as matter hidden away in Alpine "ovens," underground gemstone veins, cellphone carcasses, and burial sites. At these sources, the work of refinement is distinctively difficult, drawing attention to the relationship between certain minerals and the exceptional qualities of those who seek, circulate, and appreciate them. Were gold, gemstones, and Alpine crystals as easily sourced as the "pebbles" of Ponge's poetry (cited in the introduction), these minerals' contributions to reckoning the skills, status, taste, trustworthiness, devotion, and other qualities of the people who engage with them would be greatly diminished. Remaining hidden until uncovered by experts, the minerals found inside of discarded electronics and the spectacular crystal specimens found in Alpine "ovens" work especially well in narratives that configure "treasure hunting" as a particular sort of skilled practice that demands not only technical knowhow but also a perspective discerning enough to perceive hidden treasure. For some, then, the pursuit of precious minerals offers distinctive possibilities for self-fashioning, allowing Alpine crystal hunters' pursuit of "a lifestyle based on freedom, wilderness, discovery, dreams, adventure, and risk" (chapter three), for example, or cellphone scrappers pursuit of freedom and the American dream (chapter one).

While the preciousness of precious minerals is unquestionably associated with certain features of certain minerals, it can only be realized through multiple, engaging processes involving the correspondence of people and responsive matter, albeit sometimes in ways that can conceal, elide, or, to borrow Calvão's metaphor from chapter five, render such processes opaque (see also Ferry, 2013). Chapters by Bell, Walsh, and Vallard all draw on Ingold's, Paxson's and others' discussions of artisanal engagement with the material. In describing this engagement, Vallard in chapter four points to the necessary sensory openness of each practitioner to the proprieties of the matter at hand and also to the ways in which material "dynamically shapes" the mind and body of artisans (Naji & Douny, 2009: 418). The entangled nature of mineral and mineral work (lapidary, scrapper, artisanal miner) underwrites human-mineral interactions and generates mineral qualities, including the quality of preciousness. Such work also creates niches for specialists with unique sets of skills developed through extensive experience, as evident in Vallard's description of the work of the lapidarists who cut the Golden Jubilee Diamond and that of archaeologists conducting forensic research on the chaîne opératoire of beads. In some cases, such work is public and unabashedly messy. In Bell's chapter one, for example, dirt and demolition are key components of working with precious minerals, indexically linked to ideas of passion, joy, freedom, and "Treasure Hunting in Trash!!!!" Similarly, Calvão in chapter five notes how, in negotiations over diamonds in Angola, dirt under a miner/seller's fingernails assures buyers of the quality of what is being offered. In other cases, however, minerals' previous engagements with people are not perceptible nor are they meant to be so, as evident in the secretive processing of sapphires in Thai labs referenced by Walsh in chapter two or in Calvão's description of how cutting and polishing can "purify" diamonds by rendering them untraceable to potentially problematic sources.

As suggested in Raveneau's chapter three and Vallard's chapter four, moments of discovery or revelation can be powerful experiences for people encountering minerals, but the intimacy that attends human-mineral engagements can develop over time as well. Such is the case, for example, in the relationships that develop among Thai collectors and the hard stone beads they collect. Vallard describes how the beads that are kept and worn have profoundly inspiring affective qualities; for one collector, his beads "bind him to the entrails of earth and to a temporality before humanity, bringing him closer to an immutable nature." It is not only as affective objects of admiration that precious minerals can act, however. Elsewhere in her chapter, Vallard describes how the Emerald Buddha manifests preciousness as a non-semiotic quality in that it is "not considered to be the residence of the spirit but the spirit itself, the spirit and the substance being totally inseparable." In this case, the stone does not stand for the deity in a semiotic relation; it is the deity. This non-semiotic claim can also be found in relation with other precious substances, notably gold (Ferry, 2016), in ways that demand that we take a cautious approach to attributions of inherent agency. In chapter six, Field critiques the way in which framing gold as an actor and protagonist of its own story can reinscribe the power relations that make its value appear to be natural and inevitable. Field draws on the work of Nicholas Saunders to limn a general Andean fascination with brilliance in which gold played only a secondary role, revealing how the preciousness of this particular substance has never been inevitable.

The preciousness that emerges from intimate human-mineral engagements is often inspired by the passions that people bring to them. This passion is perhaps most obvious in people's engagements with the Emerald Buddha, with patrimonial beads or Alpine crystals, or with other singular specimens that, as noted earlier, embody in their being, more than they represent, the qualities that make them precious and that people seek out in them. Even the most intimate moments of human-mineral engagement are enabled by social scaffolding, however, involving groups of people connected through networks in which ideologies, semiotic systems, information, and skill sharing are mediated in a variety of ways: online postings of YouTube videos and their commenters (chapter one), anecdotes of death-defying or deadly climbs (chapter three), communities of practice (chapter four), and so on. As Bell and Calvão note in chapters one and five, respectively, what is shared in such social networks is valuable only as long as it is credible, meaning that the people involved commonly devote heightened attention not just to minerals but to one another as well – attention discerning enough to see what's implied by Scrapper Girl's unscarred hands (chapter one), for example. Here again, though, the distinctive qualities of particular minerals play a role in shaping the social networks through which they and (mis)information about them pass. Walsh, for example, discusses in chapter two how sapphires help shape the exchange relationships that develop around them, their one-of-a-kind nature and always ambiguous value giving force to the rules of the Geertzian bazaar in which exchange partners relate to one another as "intimate antagonists" (Geertz, 1979: 225), standing to gain not just from what they have to trade but from what they know and others do not. As Calvão notes in chapter five, the global diamond industry similarly depends not just on diamonds and their mining, processing, and marketing but also on "an active withholding of knowledge," in particular concerning potentially problematic origins.

As Calvão (chapter five) notes is the case with diamonds, all minerals circulating in the world today "have to be found, mined, and transported from somewhere," and the aforementioned scarcity and inaccessibility of certain of them is especially apparent in the risks taken and dangers faced by so many who pursue them. Malagasy artisanal miners, for example, engage in skilled, artisanal work in violation of conservation regulations, working under threat of being caught and fleeced by police and facing the ever-present possibility of dying in a collapsing shaft. Similarly, while motivated more by passion than possible fortune, the crystal hunters, described in Raveneau's chapter three, risk a great deal in their Alpine forays, following dangerous paths and seeking out the crumbly walls and rotten rocks that even sporting mountaineers avoid. Taking a Kopytoffian "cultural biography" approach, Raveneau describes how crystal hunters mobilize several forms of value (monetary, affective, symbolic, agonistic, and so on) to increase their own social prestige through crystals. His discussion of the role of risk-taking and death in the enhancement of value is especially striking; he writes: "The kind of death involved in the structure of social relations ... establishes a price ... [I]t is the assessment by others of the genuineness of the subject's commitment to his action." The centrality of death and the risk of death in the constitution of prestige as a value in crystal hunting give an immediacy and relevancy sometimes missing from the cultural biography approach, which can feel narrowly methodological. In chapter four, Vallard shows another dimension of the cultural biography approach, focusing on what happens when minerals are diverted from their paths. She recounts the story of a group of local "bead keepers" who collect antique hard stone beads and wear them around their necks. These bead keepers see themselves as having a more legitimate, local claim on the beads than the archaeologists, who will facilitate the beads' movement to museums and private collections in Bangkok. She argues that the bead keepers' actions to divert beads from this path constitute a contestation of national authority, supporting an alternative, local narrative of Thai-ness and Thai history.

The social, affective, and signifying worlds of minerals also take place within political and economic processes. Looked at from this angle, preciousness can show up as a structure of feeling in multiple situations. For the treasure hunters described by Bell in chapter one, preciousness embodies a new kind of reward for hard work and ingenuity, one that seems as though it might be more resistant and even in a position to leverage contemporary precarity. As Bell notes, scrapping cellphones (like the artisanal mining described by Walsh and Calvão in chapters two and five) has grown alongside a decline of waged employment in the United States and elsewhere. The risky work of artisanal gemstone mining is therefore particularly appealing to marginal people in contexts offering few other opportunities for getting ahead. In the case described by Calvão in chapter five, preciousness precipitates from an intriguing tension between the physical transparency of diamonds and the opacity of their origins. Despite the Kimberley Process and related certification projects to make the sources of diamonds more visible, the industry continues to rely on the argument that most diamonds are untraceable to control supply and manage expectations, as well as to cultivate diamonds' allure. The growing challenge of synthetic diamonds makes this balance between transparency and opacity riskier and costlier for De Beers and other diamond dealers. These instances, selected from among many others in the volume, show how preciousness, as a multivalent property of minerals, emerges out of the processes of labour and governance, while also appearing to be distinctively separate from such processes.

Preciousness is therefore unquestionably often an outcome of commodity fetishism, as Field argues in chapter six. But minerals offer a distinctive case, as the human labour that is essential to their extraction, circulation, and processing is commonly distinguished from the forces at work in their original production. "Natural" sapphires and diamonds, for example, are precious to some in the way that their synthetic, lab-made alternatives are not, at least partly because of how they originate apart from human intentionality. While the process of commodity fetishism can accommodate this apartness, we would argue that it cannot entirely contain it; a residue remains outside of such an account.

In the preceding paragraphs, we have highlighted how any consideration of the preciousness of precious minerals must reckon with all that enables the wide variety of human-mineral engagements described in the previous chapters: commodity fetishism; longstanding, far-reaching, and ongoing systems of exploitation; diverse regimes of value and associated ideological and semiotic systems; the social lives of minerals and the social lives that are built upon and through them;

and so on. Such reckonings can only do so much, however, to account for the scene with which we opened this volume: a room full of delighted anthropologists and mineralogists gathered around a table, loupes and minerals in hand, taken with minerals. There we were, specialists with different backgrounds and training coming together around the question of preciousness, some of us prone to looking deeply into distinctive aspects of human life for answers, others more inclined to highlight the distinctive features of minerals themselves, but all of us similarly taken with the actual preciousness at hand. Making sense of this scene requires acknowledging an emergent quality of certain human-mineral engagements that we have described here as "preciousness."

Whatever we or others make of gold, Alpine crystal specimens, sapphires, emeralds, diamonds, or any of the specimens that we handled on that day, the potential for these and other minerals to arouse fascination, wonder, devotion, or to otherwise affect people, even if only momentarily, is undeniable. This preciousness is not an attribute of minerals themselves, nor can it be said to exist only in the eye of the beholder. Rather, the preciousness of precious minerals is realized through specific human-mineral engagements, each offering distinctive opportunities for certain minerals to affect, and for certain beholders to be affected, in powerful, motivating, inspiring, and effective ways.

REFERENCES

Brazeal, B. (2014). The fetish and the stone: A moral economy of charlatans and thieves. In P.C. Johnson (Ed.), *Spirited things: The work of "possession" in Afro-Atlantic religions* (pp. 131–54). Chicago: University of Chicago Press.

Brazeal, B. (2017). Austerity, luxury and uncertainty in the Indian emerald trade. *Journal of Material Culture*, 22(4), 437–52. https://doi.org/10.1177/1359183517715809

Collier, J.S., & Ong, A. (2005). Global assemblages, anthropological problems. In A. Ong & J.S. Collier (Eds.), *Global assemblages: Technology, politics, and ethics as anthropological problems* (pp. 3–21). Malden, MA: Blackwell Publishing.

Ferry, E. (2013). *Minerals, collecting, and value across the U.S.-Mexican border.* Bloomington: University of Indiana Press.

Ferry, E. (2016). On not being a sign: Gold's semiotic claims. *Signs and society*, 4(1), 57–79. https://doi.org/10.1086/685055

Geertz, C. (1979). Suq: The bazaar economy in Sefrou. In C. Geertz, H. Geertz, and L. Rosen, *Meaning and order in Moroccan society: Three essays in cultural analysis* (pp. 123–313). Cambridge: Cambridge University Press.

Ingold, T. (2012). Toward an ecology of materials. *Annual Review of Anthropology*, *41*(1), 427–42. https://doi.org/10.1146/annurev-anthro-081309-145920

Karpik, L. (2010). *Valuing the unique: The economics of singularities.* Princeton, NJ: Princeton University Press.

Kunz, G.F. (1971). *The curious lore of precious stones: Being a description of their sentiments and folk lore, superstitions, symbolism, mysticism, use in medicine, protection, prevention, religion, and divination, crystal gazing, birthstones, lucky stones and talismans, astral, zodiacal, and planetary.* New York: Dover Publications. (Original work published 1913 by J.B. Lippincott Co.)

Naji, M., & Douny, L. (2009). Editorial. *Journal of Material Culture*, *14*(4), 411–32. https://doi.org/10.1177/1359183509346184

Naylor, T. (2010). The underworld of gemstones. *Crime, Law and Social Change*, *53*(2), 131–58. https://doi.org/10.1007/s10611-009-9223-z

Ponge, F. (1972). The pebble. B. Archer (Trans.). In *The voice of things* (pp. 69–77). New York: McGraw Hill. (Original work published 1942: Ponge, F. *Le parti pris des choses.* Paris: Gallimard)

Reno, J. (2016). *Waste away: Working and living with a North American landfill.* Oakland: University of California Press.

Tsing A.L. (2015). *The mushroom at the end of the world: On the possibility of life in capitalist ruins.* Princeton, NJ: Princeton University Press.

Contributors

Joshua A. Bell is a curator of globalization and a member of the Department of Anthropology at the Smithsonian Institution. Recent research projects focus on cellular phones as material culture and on the anthropology of travel, expeditions, and photography.

Filipe Calvão is a member of the Department of Anthropology and the Sociology of Development at the Graduate Institute of International and Development Studies in Geneva. His recent research focuses on diamond mining and trading in Angola and on gemstone auctions in Europe.

Elizabeth Ferry is a member of the Department of Anthropology at Brandeis University. Her recent research focuses on the mining, collecting, and global circulation of mineral specimens, silver and gold, and on the intersections of mining and finance.

Les W. Field is a member of the Department of Anthropology at the University of New Mexico. His research has focused on artisanal labour and political anthropology in Latin America and on the value of gold and other precious materials in historical and contemporary contexts.

Susan D. Gillespie is a member of the Department of Anthropology at the University of Florida. She is an archaeologist whose research has recently focused on the significance of jade in Mesoamerica and on human entanglement with material culture.

Gilles Raveneau is maître de conférences at Université Paris Ouest Nanterre La Défense in France. His recent research focuses on mountaineering and on the hunting and collecting of mineral specimens from the Alps and the Himalayas.

Annabel Vallard is a member of the Centre national de la recherche scientifique (CNRS, Center for Southeast Asian Studies) at L'École des hautes études en sciences sociales (EHESS), Paris. Her research focuses on the phenomenological and technical aspects of human-material (gemstones and silk) relations in Southeast Asia (mainly Thailand) and Japan.

Andrew Walsh is a member of the Department of Anthropology at the University of Western Ontario. His recent research focuses on a post-boom artisanal gemstone mining community in northern Madagascar and on the ethics of do-it-yourself (DIY) transnational humanitarian, development, and conservation projects.

Index

Milton Keynes UK
Ingram Content Group UK Ltd.
UKHW011112210424
441408UK00004B/124/J

9 781487 503178